Lecture Notes in Computer Science 6479

Commenced Publication in 1973
Founding and Former Series Editors:
Gerhard Goos, Juris Hartmanis, and Jan van Leeuwen

W0036952

Katsuhisa Horimoto   Masahiko Nakatsui
Nikolaj Popov (Eds.)

Algebraic and Numeric Biology

4th International Conference, ANB 2010
Hagenberg, Austria, July 31–August 2, 2010
Revised Selected Papers

 Springer

Volume Editors

Katsuhisa Horimoto
Masahiko Nakatsui
National Institute of Advanced Industrial Science and Technology (AIST)
Computational Biology Research Centre (CBRC)
Tokyo 135-0064, Japan
E-mail:{k.horimoto, m.nakatsui}@aist.go.jp

Nikolaj Popov
Johannes Kepler University of Linz
Research Institute for Symbolic Computation (RISC)
4040 Linz, Austria
E-mail: popov@risc.uni-linz.ac.at

ISSN 0302-9743   e-ISSN 1611-3349
ISBN 978-3-642-28066-5   ISBN 978-3-642-28067-2   (eBook)
DOI 10.1007/978-3-642-28067-2
Springer Heidelberg Dordrecht London New York

Library of Congress Control Number: 2011945509

CR Subject Classification (1998): F.3.1, F.4, D.2.4, I.1, J.3

LNCS Sublibrary: SL 1 – Theoretical Computer Science and General Issues

Typesetting: Camera-ready by author, data conversion by Scientific Publishing Services, Chennai, India

Printed on acid-free paper

Springer is part of Springer Science+Business Media (www.springer.com)

Preface

This volume contains the proceedings of the International Conference on Algebraic and Numeric Biology (ANB 2010). It was held during July 31–August 2, 2010 in the Castle of Hagenberg, Austria, and was organized by the Research Institute for Symbolic Computation (RISC) of the Johannes Kepler University, Linz, Austria, together with the National Institute of Advanced Industrial Science and Technology (AIST), Tokyo, Japan.

Algebraic and numeric biology is the interdisciplinary forum for the presentation of research on all aspects of the application of symbolic and numeric computation in biology. Algebraic and Numeric Biology was renamed after our traditional name, Algebraic Biology, to consider its wider scope of mathematical methods, especially numeric computation.

The Algebraic and Numeric Biology conference is a follow-up of the Algebraic Biology conference. Since 2005, we have been organizing an international conference, Algebraic Biology, that focuses on the application of computer algebra, automated reasoning, hybrid algebraic and numeric computation, which is also called symbolic computation, to all types of problems from biology. The first conference, AB 2005, was held in November 2005 in Tokyo, Japan. The second conference, AB 2007, was held in July 2007 in Hagenberg, Austria. The third conference, AB 2008, was held in July and August again in Hagenberg, Austria. These conferences were quite successful. For this issue of the conference, we extended the range of mathematical methods.

The initiation of the series of these conferences was motivated by the recent trends in symbolic and numeric computation and biology: In symbolic and numeric computation, the recent advances in computer performance and algorithmic methods have accelerated the extension of the scientific fields to which symbolic and numeric computation can be applied. In biology, the determination of complete genomic sequences and the subsequent improvements of experimental techniques have yielded large amounts of information about the biological molecules underlying various biological phenomena. Under these circumstances, the marriage of symbolic and numeric computation and biology is expected to generate new mathematical models for biological phenomena and new symbolic and numeric techniques for biological data analysis.

We received submissions from 16 countries (Austria, Belgium, Chile, France, Germany, Hong Kong, Ireland, Italy, Japan, Mexico, Portugal, Singapore, Spain, UK, Ukraine, and USA) and 10 papers were accepted for presentation at the conference. Each submission was assigned to at least two Program Committee members, who carefully reviewed the papers, in some cases with the help of external referees. The merits of the submissions were discussed by the Steering Committee at the dedicated meeting in Hagenberg.

We are pleased to continue our collaboration with Springer, who agreed to publish the conference proceedings in the *Lecture Notes in Computer Science* series.

We, the ANB 2010 organizers of the conference, are grateful to the following sponsors for their financial contributions toward its operation and success: the Doctoral Program Computational Mathematics at the Johannes Kepler University supported by the Austrian Science Fund (FWF), the International Studies for Informatics Hagenberg, Linzer Hochschulfonds, the National Institute of Advanced Industrial Science and Technology, and the Upper Austrian Government.

Our thanks are also due to the members of the Program Committee and to those who ensured the effective running of the conference.

July 2010

Bruno Buchberger
Katsuhisa Horimoto
Masahiko Nakatsui
Nikolaj Popov

Conference Organization

Steering Committee

Bruno Buchberger	Johannes Kepler University of Linz, Austria
Katsuhisa Horimoto	National Institute of Advanced Industrial Science and Technology, Japan
Reinhard Laubenbacher	Virginia Bioinformatics Institute, USA
Bud Mishra	New York University, USA
Nikolaj Popov	Johannes Kepler University of Linz, Austria

Program Chairs and Proceedings Editors

Katsuhisa Horimoto	National Institute of Advanced Industrial Science and Technology, Japan
Masahiko Nakatsui	National Institute of Advanced Industrial Science and Technology, Japan
Nikolaj Popov	Johannes Kepler University of Linz, Austria

Program Committee

Tatsuya Akutsu	Kyoto University, Japan
Hirokazu Anai	Kyushu University, Japan
Niko Beerenwinkel	Swiss Federal Institute of Technology Zurich, Switzerland
Armin Biere	Johannes Kepler University of Linz, Austria
François Boulier	University of Lille I, France
Bruno Buchberger	Johannes Kepler University of Linz, Austria
Luca Cardelli	Microsoft Research, Cambridge, UK
Luonan Chen	Osaka Sangyo University, Japan
Wai-Ki Ching	University of Hong Kong, China
Kwang-Hyun Cho	Korea Advanced Institute of Science and Technology, Korea
Franck Delaplace	Evry University, France
Hoon Hong	North Carolina State University, USA
Katsuhisa Horimoto	National Institute of Advanced Industrial Science and Technology, Japan
Peter Huggins	Carnegie Mellon University, USA
Abdul Salam Jarrah	Virginia Bioinformatics Institute, USA
Erich Kaltofen	North Carolina State University, USA
Hans Kestler	University of Ulm, Germany
Temur Kutsia	Johannes Kepler University of Linz, Austria

Philipp Kügler	RICAM, Austria and University of Stuttgart, Germany
Doheon Lee	Korea Advanced Institute of Science and Technology, Korea
James Lynch	Clarkson University, USA
Manfred Minimair	Seton Hall University, USA
Stefan Müller	RICAM, Austria
Masahiko Nakatsui	National Institute of Advanced Industrial Science and Technology, Japan
Masahiro Okamoto	Kyushu University Japan
Eugenio Omodeo	University of Trieste, Italy
Sonja Petrovic	University of Illinois at Chicago, USA
Nikolaj Popov	Johannes Kepler University of Linz, Austria
Georg Regensburger	RICAM, Austria
Carolyn Talcott	SRI International, USA
Ashish Tiwari	SRI International, USA
Hiroyuki Toh	Kyushu University, Japan
Dongming Wang	Beihang University, China and UPMC-CNRS, France
Limsoon Wong	National University of Singapore
Ruriko Yoshida	University of Kentucky, USA

Local Organization Chair

Nikolaj Popov	Johannes Kepler University of Linz, Austria

Table of Contents

P_0-Matrix Products of Matrices

Murad Banaji

Department of Mathematics, University of Portsmouth, Lion Gate Building,
Lion Terrace, Portsmouth, Hampshire PO1 3HF, UK

Abstract. The question of when the product of two matrices lies in the closure of the P-matrices is discussed. Both sufficient and necessary conditions for this to occur are derived. Such results are applicable to questions on the injectivity of functions, and consequently the possibility of multiple fixed points of maps and flows. General results and special cases are presented, and the concepts illustrated with numerous examples. Graph-theoretic corollaries to the matrix-theoretic results are touched upon.

Keywords: P-matrix; P_0-matrix; matrix factorisation; injectivity; stability.

1 Introduction

We will be interested in P-matrices and certain closely related classes of matrices. In each of the definitions below, A is a real square matrix:

Definition 1. *A is a P-matrix if all of its principal minors are positive.*

Definition 2. *A is a P_0-matrix if all of its principal minors are nonnegative.*

Definition 3. *A is a $P^{(-)}$-matrix if $-A$ is a P-matrix.*

Definition 4. *A is $P_0^{(-)}$-matrix if $-A$ is a P_0-matrix.*

Matrices belonging to any of these classes will be termed *P-type matrices*. P-type matrices play important roles in various applications, including to the linear complementarity problem [1], and to a variety of questions in biology [2].

This paper explores necessary and sufficient conditions for the product of two matrices to be a P_0-matrix. The work is motivated primarily (but not solely) by two observations:

1. Differentiable functions with P-matrix Jacobians on certain domains are injective [3]. The same is true for functions with P_0-matrix Jacobians, provided certain additional conditions are satisfied [4]. Such results have been applied to problems in economics [5], biology [6] and chemistry [7].
2. Matrices which arise as Jacobian matrices in various areas of application often have natural factorisations as a consequence of physical constraints [7,8]. In practice, it is often the case that some of the factors are constant, while only the sign pattern of others is known. Determining whether all allowed products are P-type matrices is a nontrivial problem.

K. Horimoto, M. Nakatsu, and N. Popov (Eds.): ANB 2011, LNCS 6479, pp. 1–17, 2012.
© Springer-Verlag Berlin Heidelberg 2012

Here, rather than focussing on the applications, a number of matrix-theoretic results which have proved useful are reviewed and generalised, while previously published applications are referenced. The results are in most cases amenable to straightforward algorithmic implementation. Elementary proofs are presented and some graph-theoretic corollaries are touched upon.

1.1   Some Properties of P-Type Matrices

It follows from elementary properties of P-matrices that any P_0-matrix plus a positive diagonal matrix is a P-matrix. In fact, the P_0-matrices can be characterised as follows:

Lemma 1. *A square matrix A is a P_0-matrix if and only if $\det(A + D) > 0$ for every positive diagonal matrix D.*

Proof. See Section 3.2 in [8].   □

Eigenvalues and Stability. Other properties of P-matrices include that their eigenvalues are excluded from a certain wedge around the negative real axis [9], implying that real eigenvalues of P-matrices are positive, and further that 2×2 P-matrices are **positive stable** (i.e., have eigenvalues with positive real part). Positive stability for P-matrices satisfying additional constraints can also be deduced in higher dimensions, and a result in this direction of Hershkowitz and Keller (Theorem 1.1 in [10]) will be applied later to prove that certain matrices are positive stable.

Injectivity of Functions. Define a **rectangular domain** in $\mathbb{R}^n$ to be the product of intervals on the coordinate axes, each of which may be open, closed or semi-open. The following result of Gale and Nikaido has proved highly applicable: given a rectangular domain $X \subseteq \mathbb{R}^n$, each differentiable function $f : X \to \mathbb{R}^n$ with P-matrix Jacobian is injective on X (Theorem 4 and subsequent remarks in [3]). If X is, additionally, open, then it suffices for the Jacobian matrix to be a nonsingular P_0-matrix (Theorem 4w in [3], and see also [4]). For simplicity, the results presented below involve P- and P_0-matrices, but, since injectivity of $-f$ is equivalent to injectivity of f, all results have easy duals stated in terms of $P^{(-)}$- and $P_0^{(-)}$-matrices. It is the dual results which generally arise in practice, and most of the previous results referenced actually involve $P^{(-)}$- and $P_0^{(-)}$-matrices. This point will not be laboured.

1.2   The Main Question

Notation. Let $\mathbb{R}^{n \times m}$ denote the real $n \times m$ matrices. Define $\mathcal{P}_0$ to be the set of P_0-matrices, and $\mathcal{P}_{0,n} \subseteq \mathbb{R}^{n \times n}$ to be the set of $n \times n$ P_0-matrices. Clearly, $\mathcal{P}_{0,n}$ is the closure of the set of $n \times n$ P-matrices.

P_0-**matrix Products.** It has been well illustrated in previous work that matrices which arise as Jacobian matrices in applications often have a natural factorisation [11,8]. Further, by Lemma 1, showing that $A \in \mathcal{P}_0$ is both necessary and sufficient to ensure that $A + D$ is nonsingular (and is in fact a P-matrix) for each positive diagonal matrix D. This is important because in certain classes of applications, Jacobian matrices can be written as such a sum with the magnitude of entries in D unknown, and thus proving that the first matrix A is a P_0-matrix becomes both necessary and sufficient to rule out singularity of the Jacobian matrix (which in turn is sufficient to rule out saddle-node bifurcations).

The key question in this paper is most generally phrased in terms of matrix-valued functions. Given an arbitrary set X, let $\mathcal{A} : X \to \mathbb{R}^{n \times m}$ be some function ascribing to each $x \in X$ an $n \times m$ matrix $\mathcal{A}(x)$. Similarly, let $\mathcal{B} : X \to \mathbb{R}^{m \times n}$ ascribe to each $x \in X$ the $m \times n$ matrix $\mathcal{B}(x)$. The question is:

   When is $\{\mathcal{A}(x)\mathcal{B}(x) \,|\, x \in X\} \subset \mathcal{P}_0$?

We will answer this question for a number of cases of importance. In applications, X is often the state space of some dynamical system, assumed to be a rectangular subset of $\mathbb{R}^n$, for reasons touched on above. However, this restriction on X is unnecessary for the matrix-theoretic results presented here.

2   Notation and Definitions

Sets of Real Numbers. Define $\mathbb{R}_{>0} \equiv (0, \infty)$, $\mathbb{R}_{\geq 0} \equiv [0, \infty)$, $\mathbb{R}_{<0} \equiv (-\infty, 0)$ and $\mathbb{R}_{\leq 0} \equiv (-\infty, 0]$. A set of real numbers $\mathcal{R}$ is **signed** if $\mathcal{R} \subseteq \mathbb{R}_{>0}$ or $\mathcal{R} \subseteq \mathbb{R}_{<0}$ or $\mathcal{R} = \{0\}$, and is **weakly signed** if $\mathcal{R} \subseteq \mathbb{R}_{\geq 0}$ or $\mathcal{R} \subseteq \mathbb{R}_{\leq 0}$. A set of real numbers which fails to be weakly signed (i.e., which intersects both $\mathbb{R}_{>0}$ and $\mathbb{R}_{<0}$) is **unsigned**.

Classes of Matrices. P-type matrices have already been defined above. Real matrices all of whose eigenvalues have negative real part are said to be **Hurwitz**: a square matrix M is Hurwitz if and only if $-M$ is positive stable. $\mathcal{D}_n$ will refer to the set of $n \times n$ diagonal matrices with positive diagonal entries.

Submatrices, Minors, etc. For any positive integer n, define $\mathcal{I}_n = \{1, \ldots, n\}$. Let A be an $n \times m$ matrix, with $\alpha \subseteq \mathcal{I}_n$, $\beta \subseteq \mathcal{I}_m$ nonempty. The following notation will be used:

- $A(\alpha|\beta)$ is the submatrix of A with rows indexed by α and columns indexed by β.
- $A[\alpha|\beta] \equiv \det(A(\alpha|\beta))$. We write $A[\alpha]$ as shorthand for $A[\alpha|\alpha]$.
- $A_{\alpha\beta}$ is an $n \times m$ matrix defined by $(A_{\alpha\beta})_{ij} = A_{ij}$ if $i \in \alpha$ and $j \in \beta$ and $(A_{\alpha\beta})_{ij} = 0$ otherwise.

Consider some minor $A_{\alpha\beta}[\gamma|\delta]$. If $|\gamma| = |\alpha|$, then either $\gamma = \alpha$, or $A_{\alpha\beta}(\gamma|\delta)$ contains a row of zeros; similarly If $|\delta| = |\beta|$, then either $\delta = \beta$, or $A_{\alpha\beta}(\gamma|\delta)$ contains a column of zeros. In particular, all $|\alpha| \times |\beta|$ submatrices of $A_{\alpha\beta}$, apart possibly from $A(\alpha|\beta)$, must contain a row or column of zeros (and hence, if they are square, must be identically singular).

Example 1. Given the matrix:

$$A = \begin{pmatrix} 2 & 1 & 1 \\ 1 & 1 & 1 \\ 0 & 3 & 1 \end{pmatrix},$$

we have

$$A(\{1,2\}|\{1,3\}) = \begin{pmatrix} 2 & 1 \\ 1 & 1 \end{pmatrix}, \quad A[\{1,2\}|\{1,3\}] = 1, \quad \text{and} \quad A_{\{1,2\}\{1,3\}} \begin{pmatrix} 2 & 0 & 1 \\ 1 & 0 & 1 \\ 0 & 0 & 0 \end{pmatrix}.$$

Matrix-Valued Functions, Sets of Matrices and Independence. Given a set X and matrix-valued functions $\mathcal{A} : X \to \mathbb{R}^{n \times m}$, $\mathcal{B} : X \to \mathbb{R}^{m \times n}$. Define $\mathcal{A}(X) = \{\mathcal{A}(x) : x \in X\}$, $\mathcal{B}(X) = \{\mathcal{B}(x) \,|\, x \in X\}$, and $(\mathcal{AB})(X) = \{\mathcal{A}(x)\mathcal{B}(x) \,|\, x \in X\}$. Naturally, $\mathcal{A}^T : X \to \mathbb{R}^{m \times n}$ will be defined by $\mathcal{A}^T(x) = [\mathcal{A}(x)]^T$, $\mathcal{A}_{ij} : X \to \mathbb{R}$ will be defined by $\mathcal{A}_{ij}(x) = [\mathcal{A}(x)]_{ij}$. Similarly, the functions $\mathcal{A}(\delta|\gamma) : X \to \mathbb{R}^{|\delta| \times |\gamma|}$, $\mathcal{A}[\delta|\gamma] : X \to \mathbb{R}$, $\mathcal{A}_{\delta\gamma} : X \to \mathbb{R}^{n \times m}$, etc. all have their natural meanings. From here on, X will always refer to an arbitrary set, which is the domain of some matrix-valued functions.

If, for each $A \in \mathcal{A}(X)$, $B \in \mathcal{B}(X)$, there exists $x \in X$ such that $A = \mathcal{A}(x)$, $B = \mathcal{B}(x)$, then we will say that $\mathcal{A}$ and $\mathcal{B}$ are **independent**. Observe that for independent functions, $(\mathcal{AB})(X) = \{AB \,|\, A \in \mathcal{A}(X), B \in \mathcal{B}(X)\}$. Where $\mathcal{A}$ and $\mathcal{B}$ are defined as sets of $n \times m$ and $m \times n$ matrices respectively (rather than matrix-valued functions), the set $\mathcal{AB} = \{AB \,|\, A \in \mathcal{A}, B \in \mathcal{B}\}$ can always be regarded as the image of the product of independent matrix-valued functions (defined, for example, as projections on the Cartesian product $\mathcal{A} \times \mathcal{B}$). For brevity we will refer to $\mathcal{AB}$ as an **independent product**.

Qualitative Classes. A matrix $A \in \mathbb{R}^{n \times m}$ determines a qualitative class $\mathcal{Q}(A) \subset \mathbb{R}^{n \times m}$ [12,13], the set of all matrices with the same dimensions and sign pattern as A. $\mathcal{Q}(A)$ is convex and hence path connected, and so, by continuity of the determinant, all matrices in $\mathcal{Q}(A)$ are nonsingular if and only if either they all have positive determinant, or they all have negative determinant. Define $\mathcal{Q}_0(A)$ to be the closure of $\mathcal{Q}(A)$.

Sign-Classes. A set $\mathcal{A}$ of $n \times m$ matrices is termed a left sign-class if $A_{\alpha\mathcal{I}_m} \in$ cl($\mathcal{A}$) for each $A \in \mathcal{A}$, $\alpha \subseteq \mathcal{I}_n$. $\mathcal{A}$ is a right sign-class if $A_{\mathcal{I}_n\beta} \in$ cl($\mathcal{A}$) for each $A \in \mathcal{A}$, $\beta \subseteq \mathcal{I}_m$. If $A_{\alpha\beta} \in$ cl($\mathcal{A}$) for each $A \in \mathcal{A}$, $\alpha \subseteq \mathcal{I}_n$, $\beta \subseteq \mathcal{I}_m$, then $\mathcal{A}$ is a sign-class. More intuitively, taking any matrix in a left sign-class $\mathcal{A}$ and setting an arbitrary subset of rows to zero gives a matrix in the closure of $\mathcal{A}$. Similarly, if $\mathcal{A}$ is a right sign-class, then setting an arbitrary subset of columns to zero gives a matrix in the closure of $\mathcal{A}$. If $\mathcal{A}$ is a sign-class, then simultaneously setting some subset of rows to zero, and some subset of columns to zero, gives a matrix in the closure of $\mathcal{A}$. It is not hard to see that $\mathcal{A}$ is a sign-class if and only if it is both a left sign-class and a right sign-class. Qualitative classes are examples of sign-classes, but sign-classes need not be qualitative classes (see Example 2 below). A function $\mathcal{A} : X \to \mathbb{R}^{n \times m}$ will be termed a sign-class (resp. left sign-class, resp. right sign-class) if $\mathcal{A}(X)$ is such.

Example 2. Given an $n \times m$ matrix A, the sets

$$\{DA \,|\, D \in \mathcal{D}_n\}, \ \{AD \,|\, D \in \mathcal{D}_m\} \text{ and } \{D_1 A D_2 \,|\, D_1 \in \mathcal{D}_n, D_2 \in \mathcal{D}_m\}$$

are a left sign-class, a right sign-class and a sign-class respectively. Thus for example the set of matrices

$$\left\{ \begin{pmatrix} ac & ad \\ bc & bd \end{pmatrix}, \quad a, b, c, d > 0 \right\},$$

which can be factorised via

$$\begin{pmatrix} ac & ad \\ bc & bd \end{pmatrix} = \begin{pmatrix} a & 0 \\ 0 & b \end{pmatrix} \begin{pmatrix} 1 & 1 \\ 1 & 1 \end{pmatrix} \begin{pmatrix} c & 0 \\ 0 & d \end{pmatrix},$$

is a sign-class. Note, however, that this set of matrices is not a qualitative class: not every 2×2 matrix with positive entries can be factorised in this way.

Compatible Matrices. Given two $n \times n$ matrices A, B, define $\sigma(A, B) = \det(A)\det(B)$. Given $n \times m$ matrices A, B, the pair (A, B) will be termed **compatible** if for each $\alpha \subseteq \mathcal{I}_n, \beta \subseteq \mathcal{I}_m$ with $|\alpha| = |\beta|$, $\sigma(A[\alpha|\beta], B[\alpha|\beta]) \geq 0$. The definition extends naturally to the functions $\mathcal{A} : X \to \mathbb{R}^{n \times m}$ and $\mathcal{B} : X \to \mathbb{R}^{n \times m}$: $\mathcal{A}$ and $\mathcal{B}$ are compatible if $\mathcal{A}(x)$ and $\mathcal{B}(x)$ are compatible for each $x \in X$. From the definition, if $\mathcal{A}$ and $\mathcal{B}$ are independent, then they are compatible if and only if for each $\alpha \subseteq \mathcal{I}_n, \beta \subseteq \mathcal{I}_m$, either:

1. $\mathcal{A}[\alpha|\beta](X) = \{0\}$, or
2. $\mathcal{B}[\alpha|\beta](X) = \{0\}$, or
3. $\mathcal{A}[\alpha|\beta](X)$ and $\mathcal{B}[\alpha|\beta](X)$ are both weakly signed with either $\mathcal{A}[\alpha|\beta](X)$, $\mathcal{B}[\alpha|\beta](X) \subseteq \mathbb{R}_{\geq 0}$ or $\mathcal{A}[\alpha|\beta](X), \mathcal{B}[\alpha|\beta](X) \subseteq \mathbb{R}_{\leq 0}$.

Example 3. The matrices:

$$A = \begin{pmatrix} 2 & 0 & 1 \\ 1 & 1 & 1 \end{pmatrix} \quad \text{and} \quad B = \begin{pmatrix} 1 & 0 & 1 \\ 2 & 1 & 1 \end{pmatrix}$$

fail to be compatible because

$$A[\{1,2\}|\{1,3\}] = \begin{vmatrix} 2 & 1 \\ 1 & 1 \end{vmatrix} = 1 \quad \text{and} \quad B[\{1,2\}|\{1,3\}] = \begin{vmatrix} 1 & 1 \\ 2 & 1 \end{vmatrix} = -1$$

have opposite signs.

3   Main Results

3.1   Preliminary Results

The next three lemmas are slightly adapted from [8].

Lemma 2. *Consider an $n \times m$ matrix A and an $m \times n$ matrix B. If A and B^T are compatible, then $AB \in \mathcal{P}_{0,n}$.*

Proof. The result follows from the Cauchy-Binet formula ([14] for example) which gives, for any nonempty $\alpha \subseteq \mathcal{I}_n$:

$$(AB)[\alpha] = \sum_{\substack{\beta \subseteq \mathcal{I}_m \\ |\beta|=|\alpha|}} A[\alpha|\beta]B[\beta|\alpha].$$

By compatibility, of A and B^T, $A[\alpha|\beta]B[\beta|\alpha] \geq 0$, and thus $(AB)[\alpha] \geq 0$. Since α was arbitrary, this proves that AB is a P_0-matrix. $\square$

Clearly, compatibility of matrices A and B^T is not necessary for AB to be a P_0-matrix: for example, the matrices A and B in Example 3 fail to be compatible, but AB^T is a P-matrix. However, given two sets of matrices, compatibility may be necessary to ensure that all products are P_0-matrices:

Lemma 3. *Consider sets of matrices $\mathcal{A} \subseteq \mathbb{R}^{n \times m}$ and $\mathcal{B} \subseteq \mathbb{R}^{m \times n}$. Define the independent product $\mathcal{A}\mathcal{B}$. Assume that, $\mathcal{A}\mathcal{B} \subseteq \mathcal{P}_{0,n}$, and either:*

1. $\mathcal{B}$ is a left sign-class, or
2. $\mathcal{A}$ is a right sign-class.

Then $\mathcal{A}$ and $\mathcal{B}^T$ are compatible.

Proof. The case where $\mathcal{B}$ is a left sign-class is treated; the other case is similar. Suppose that $\mathcal{A}$ and $\mathcal{B}^T$ are not compatible, i.e., there are $\alpha \subseteq \mathcal{I}_n$, $\beta \subseteq \mathcal{I}_m$ such that $A \in \mathcal{A}$ and $B \in \mathcal{B}$ satisfy $A[\alpha|\beta]B[\beta|\alpha] < 0$. Since $\mathcal{B}$ is a left sign-class, $B_{\beta\mathcal{I}_n} \in \mathrm{cl}(\mathcal{B})$. Moreover, $(AB_{\beta\mathcal{I}_n})[\alpha] = A[\alpha|\beta]B[\beta|\alpha] < 0$ (since $B_{\beta\mathcal{I}_n}[\beta|\alpha] = B[\beta|\alpha]$, and all minors $B_{\beta\mathcal{I}_n}[\beta'|\alpha]$ with $\beta' \neq \beta$ are zero). Thus $AB_{\beta\mathcal{I}_n} \notin \mathcal{P}_{0,n}$. Since $\mathcal{P}_{0,n}$ is closed, its complement is open. Thus any matrix sufficiently near to $AB_{\beta\mathcal{I}_n}$ fails to be a P_0-matrix. By independence of $\mathcal{A}$ and $\mathcal{B}$, $AB_{\beta\mathcal{I}_n} \in \mathrm{cl}(\mathcal{A}\mathcal{B})$, and so there are matrices in $\mathcal{A}\mathcal{B}$ which fail to be P_0-matrices. The argument is similar if $\mathcal{A}$ is a right sign-class. $\square$

Example 4. Consider the matrices:

$$A = \begin{pmatrix} 2 & 1 & 0 \\ 1 & 1 & 1 \\ 1 & 3 & 1 \end{pmatrix} \quad \text{and} \quad B = \begin{pmatrix} 1 & 2 & 0 \\ 1 & 1 & 1 \\ 1 & 3 & 1 \end{pmatrix}.$$

We can check that AB is a P-matrix. However A and B^T are not compatible, as $A[\{1,2\}|\{1,2\}]B[\{1,2\}|\{1,2\}] < 0$. By Example 2, $\{DB \mid D \in \mathcal{D}_3\}$ is a left sign-class. Consequently, by Lemma 3, there exists $D \in \mathcal{D}_3$ such that ADB fails to be a P_0-matrix. For example, it can be checked that ADB is not a P_0-matrix if we choose

$$D = \begin{pmatrix} 5 & 0 & 0 \\ 0 & 5 & 0 \\ 0 & 0 & 1 \end{pmatrix}.$$

Lemma 4. *Consider the functions* $\mathcal{A} : X \to \mathbb{R}^{n \times m}$ *and* $\mathcal{B} : X \to \mathbb{R}^{m \times n}$. *Assume that* $\mathcal{A}$ *and* $\mathcal{B}$ *are independent, one of* $\mathcal{A}$ *or* $\mathcal{B}$ *is a sign-class, and* $(\mathcal{AB})(X) \subseteq P_{0,n}$. *Then* $\mathcal{A}$ *and* $\mathcal{B}^T$ *are compatible.*

Proof. By assumption, either $\mathcal{A}$ must be a right sign-class or $\mathcal{B}$ must be a left sign-class. The result now follows immediately from Lemma 3. □

Lemmas 2, 3 and 4 are the basic results which can be specialised to give necessary and sufficient conditions for matrix-products to be P_0.

Remark 1. Where a product of more than two matrices is concerned, there is no elementary generalisation of the notion of compatibility. However, an elegant necessary and sufficient graph-theoretic condition for the product of an arbitrary number of qualitative classes to consist of P_0-matrices can be constructed [15].

3.2   Qualitative Classes and Related Ideas

In applications, it frequently arises that we know the signs of quantities, but not their magnitudes. When this applies to the entries in a matrix, then our knowledge of the matrix is only that it belongs to a particular qualitative class. From the definitions of qualitative classes:

1. If $B \in \mathcal{Q}(A)$, then $\mathcal{Q}(B) = \mathcal{Q}(A)$,
2. If $B \in \mathcal{Q}_0(A)$, then $\mathcal{Q}(B) \subseteq \mathcal{Q}_0(A)$, and so $\mathcal{Q}_0(B) \subseteq \mathcal{Q}_0(A)$.

Sign Nonsingularity and Sign Singularity. Define S_n to be the set of $n \times n$ singular matrices. Let A be some $n \times n$ matrix. Then A is sign-singular (SS) if $\mathcal{Q}(A) \subseteq S_n$. A is sign-nonsingular (SNS) [13] if $\mathcal{Q}(A) \cap S_n = \emptyset$. Note that if A is SNS, then convexity, and hence path-connectedness, of $\mathcal{Q}(A)$ implies that all matrices in $\mathcal{Q}(A)$ have determinant of the same sign. Following the terminology of [13], a square matrix which is either sign nonsingular or sign singular is a matrix with *signed determinant*. If A has signed determinant, $\mathrm{sign}(\det(\mathcal{Q}(A)))$ is well defined.

Notation. Let SNS_n be the set of $n \times n$ SNS matrices, SS_n be the set of $n \times n$ SS matrices, and S_n to be the set of $n \times n$ singular matrices.

Remark 2. Note that S_n is closed by continuity of the determinant. SNS_n and SS_n consist of entire qualitative classes, namely:

1. If $A \in \mathrm{SNS}_n$, then $\mathcal{Q}(A) \subseteq \mathrm{SNS}_n$.
2. If $A \in \mathrm{SS}_n$, then $\mathcal{Q}(A) \subseteq \mathrm{SS}_n$.

Both statements follow immediately from the definitions.

Characterising Matrices Which are SNS or SS. Given an $n \times n$ matrix A, and a permutation $\sigma = [\sigma_1, \ldots, \sigma_n]$ of the list $[1, \ldots, n]$, define the $n \times n$ matrix A^σ by $A_{ij}^\sigma = A_{ij}$ if $j = \sigma_i$, and $A_{ij}^\sigma = 0$ otherwise. Note that $A^\sigma \in \mathcal{Q}_0(A)$. The complement of the set of square matrices which are SNS or SS is characterised as follows:

Lemma 5. *Let A be an $n \times n$ matrix. Then $A \notin (\mathrm{SNS}_n \cup \mathrm{SS}_n)$ if and only if there exist $A_1, A_2 \in \mathcal{Q}(A)$ with $\det(A_1)\det(A_2) < 0$.*

Proof. Suppose there exist $A_1, A_2 \in \mathcal{Q}(A)$ with $\det(A_1)\det(A_2) < 0$. Clearly $A \notin \mathrm{SS}_n$. On the other hand, convexity and hence path-connectedness of $\mathcal{Q}(A)$ implies that there exists $A_3 \in \mathcal{Q}(A)$ which is singular, so $A \notin \mathrm{SNS}_n$.

Conversely, $\det(A)$ is simply a polynomial in the entries A_{ij}, and the assumption that $A \notin (\mathrm{SNS}_n \cup \mathrm{SS}_n)$ means that $\det(A)$ consists of a sum containing at least one positive monomial, corresponding say to permutation σ_1 of $[1, \ldots, n]$, and at least one negative monomial, corresponding say to permutation σ_2 of $[1, \ldots, n]$. Note that $A^{\sigma_1}, A^{\sigma_2} \in \mathcal{Q}_0(A)$ and $\det(A^{\sigma_1}) > 0$, and $\det(A^{\sigma_2}) < 0$. By continuity of the determinant, there exist matrices in $\mathcal{Q}(A)$ with both positive and negative determinants. □

Example 5. Consider the matrix

$$A = \begin{pmatrix} a & b \\ c & d \end{pmatrix}$$

where $a, b, c, d > 0$. We can confirm that $A \notin (\mathrm{SNS}_n \cup \mathrm{SS}_n)$. The matrices

$$A^{[1,2]} = \begin{pmatrix} a & 0 \\ 0 & d \end{pmatrix} \text{ and } A^{[2,1]} = \begin{pmatrix} 0 & b \\ c & 0 \end{pmatrix}$$

both lie in the closure of $\mathcal{Q}(A)$, and have determinants of opposite sign. So, for example,

$$A_1 = \begin{pmatrix} a & \epsilon \\ \epsilon & d \end{pmatrix} \text{ and } A_2 = \begin{pmatrix} \delta & b \\ c & \delta \end{pmatrix}$$

with $\epsilon = \sqrt{ad}/2$ and $\delta = \sqrt{bc}/2$ are both in $\mathcal{Q}(A)$, and have determinants of opposite sign.

Lemma 6. *Let A be an $n \times n$ matrix. Then $A \notin (\mathrm{SNS}_n \cup \mathrm{S}_n)$ if and only if there exists $B \in \mathcal{Q}(A)$ satisfying $\det(A)\det(B) < 0$.*

Proof. That $\det(A)\det(B) < 0$ implies $A \notin (\mathrm{SNS}_n \cup \mathrm{S}_n)$ is immediate. In the opposite direction, note that $A \notin (\mathrm{SNS}_n \cup \mathrm{S}_n)$ implies that $A \notin (\mathrm{SNS}_n \cup \mathrm{SS}_n)$. By Lemma 5, there certainly exist matrices in $\mathcal{Q}(A)$ with determinants of all signs. Since A is nonsingular, we simply choose B to be some matrix in $\mathcal{Q}(A)$ with determinant of opposite sign to A to get the result. □

Lemma 7. *The following hold:*

1. SS_n *is closed.*
2. *If $A \in SS_n$, then $\mathcal{Q}_0(A) \subseteq SS_n$.*
3. *If $A \in SNS_n$, then $\mathcal{Q}_0(A) \subseteq (SNS_n \cup SS_n)$.*
4. $SNS_n \cup SS_n$ *is closed*
5. $SNS_n \cup S_n$ *is closed.*

Proof. **1.** Consider a sequence $(A_i) \subset SS_n$ with $A_i \to A$. By continuity of the determinant, $A \in S_n$. Consider any $B \in \mathcal{Q}(A)$. It is easy to construct a sequence $(B_i) \subset SS_n$ with $B_i \to B$, so $B \in S_n$. (For example, choose $(B_i)_{jk} = (A_i)_{jk}$ when $A_{jk} = 0$, and $(B_i)_{jk} = B_{jk}(A_i)_{jk}/A_{jk}$ otherwise; then $B_i \in \mathcal{Q}(A_i)$, and so $B_i \in SS_n$.) Since B was arbitrary, $\mathcal{Q}(A) \subseteq S_n$, i.e., $A \in SS_n$.

2. If $A \in SS_n$, then by definition $\mathcal{Q}(A) \subseteq SS_n$. By closure of SS_n, $\mathcal{Q}_0(A) \subseteq SS_n$.

3. Let $A \in SNS_n$, and assume for definiteness that $\det(A) > 0$. Suppose there exists $B \in \mathcal{Q}_0(A) \backslash (SNS_n \cup SS_n)$. By continuity of the determinant, $\det(B) \geq 0$, and since $\mathcal{Q}(B) \subseteq \mathcal{Q}_0(A)$, $\det(C) \geq 0$ for each $C \in \mathcal{Q}(B)$. But since $B \notin (SNS_n \cup SS_n)$, by Lemma 5 there exists $C \in \mathcal{Q}(B)$ with $\det(C) < 0$, a contradiction. The argument is similar if $\det(A) < 0$.

4. By parts 2 and 3, $A \in (SNS_n \cup SS_n)$ implies that $\mathcal{Q}_0(A) \subseteq (SNS_n \cup SS_n)$. Since there are a finite number of closed qualitative classes of $n \times n$ matrices, $SNS_n \cup SS_n$ is the finite union of closed sets and is hence closed.

5. Since $SS_n \subseteq S_n$, $SNS_n \cup S_n = (SNS_n \cup SS_n) \cup S_n$. Moreover, S_n is closed by continuity of the determinant. As the union of two closed sets, $SNS_n \cup S_n$ is closed. $\qquad\square$

3.3 Compatibility of all Matrices in a Qualitative Class

Consider some matrix C. What are necessary/sufficient conditions for every pair of matrices in $\mathcal{Q}(C)$ to be compatible? Note that if every pair of matrices in $\mathcal{Q}(C)$ are compatible, then this extends by closure to $\mathcal{Q}_0(C)$.

Definition 5. *A matrix A will be termed* **completely sign determined (CSD)** *if every square submatrix of A has signed determinant. $A \in$ CSD will mean that A is a CSD matrix, and $A \in CSD_{n \times m}$ will mean that A is an $n \times m$ CSD matrix.*

The next two results show that the answer to the question posed above is precisely: "Every two matrices in $\mathcal{Q}_0(C)$ are compatible if and only if $C \in$ CSD." Consequently, since $\mathcal{Q}_0(C)$ is a sign-class, by Lemmas 2 and 4 all matrices $\{AB \mid A, B^T \in \mathcal{Q}_0(C)\}$, are P_0-matrices if and only if $C \in$ CSD.

Lemma 8. *If $C \in$ CSD, then for any $A, B \in \mathcal{Q}_0(C)$, A and B are compatible.*

Proof. Let C be an $n \times m$ matrix. Choose any $A, B \in \mathcal{Q}_0(C)$ and any $\alpha \subseteq \mathcal{I}_n$, $\beta \subseteq \mathcal{I}_m$ with $|\alpha| = |\beta|$. If $C(\alpha|\beta) \in SS_{|\alpha|}$, then $A[\alpha|\beta] = 0$ and so $A[\alpha|\beta]B[\alpha|\beta] = 0$. If $C(\alpha|\beta) \in SNS_{|\alpha|}$, then $C[\alpha|\beta] \neq 0$ and further $C[\alpha|\beta]A[\alpha|\beta] \geq 0$ and $C[\alpha|\beta]B[\alpha|\beta] \geq 0$. Thus $(C[\alpha|\beta]A[\alpha|\beta])(C[\alpha|\beta]B[\alpha|\beta]) \geq 0$. Dividing by $(C[\alpha|\beta])^2$ (which is nonzero), we have $A[\alpha|\beta]B[\alpha|\beta] \geq 0$. $\qquad\square$

Lemma 9. *Let C be an $n \times m$ matrix. If $C \notin$ CSD, then there exist $A, B \in \mathcal{Q}(C)$, such that A and B fail to be compatible.*

Proof. Since $C \notin$ CSD, there must be $\alpha \subseteq \mathcal{I}_n$, $\beta \subseteq \mathcal{I}_m$ with $|\alpha| = |\beta|$ and such that $C(\alpha|\beta) \notin (\mathrm{SNS}_n \cup \mathrm{SS}_n)$. By Lemma 5, there exist $A, B \in \mathcal{Q}(C)$, such that $A[\alpha|\beta]B[\alpha|\beta] < 0$. Thus A and B fail to be compatible. $\square$

Theorem 1. *Given a matrix C, every two matrices in $\mathcal{Q}_0(C)$ are compatible if and only if $C \in$ CSD. Consequently all matrices AB^T, where $A, B \in \mathcal{Q}_0(C)$, are P_0-matrices if and only if $C \in$ CSD.*

Proof. The first statement follows immediately from Lemmas 8 and 9. The second follows from Lemmas 2 and 4, noting that qualitative classes are sign-classes, and that and $\mathcal{Q}(C)\mathcal{Q}(C^T)$ is an independent product. $\square$

Example 6. The following matrix can be checked to be CSD:

$$\begin{pmatrix} 1 & 1 & 0 \\ -1 & 1 & -1 \\ 0 & 1 & 1 \end{pmatrix}.$$

Consequently the product of any two matrices with the sign patterns

$$\begin{pmatrix} + & + & 0 \\ - & + & - \\ 0 & + & + \end{pmatrix} \quad \text{and} \quad \begin{pmatrix} + & - & 0 \\ + & + & + \\ 0 & - & + \end{pmatrix}$$

is necessarily a P_0 matrix. Phrased differently, the product matrix

$$\begin{pmatrix} a_2b_3 + a_1b_1 & a_2b_4 - a_1b_2 & a_2b_5 \\ a_4b_3 - a_3b_1 & a_5b_6 + a_4b_4 + a_3b_2 & a_4b_5 - a_5b_7 \\ a_6b_3 & a_6b_4 - a_7b_6 & a_7b_7 + a_6b_5 \end{pmatrix}$$

is a P_0-matrix for any nonnegative values of a_i, b_i. It is important to note that the sign pattern of the product is not constant, but nevertheless all principal minors are nonnegative.

A check for whether a matrix is CSD can also be sufficient to confirm that an entire qualitative class consists of P_0-matrices:

Corollary 1. *Consider an $n \times n$ matrix A with nonnegative diagonal entries, and let I be the $n \times n$ identity matrix. If $A + I \in$ CSD, then $\mathcal{Q}_0(A) \subseteq P_{0,n}$.*

Proof. Define $C = A + I$. Then $A \in \mathcal{Q}_0(C)$ and $I \in \mathcal{Q}_0(C)$. Since $C \in$ CSD, Lemma 8 implies that A and I (and hence A and I^T) are compatible. Hence, by Lemma 2, $A(= AI^T) \in P_{0,n}$. Since $\mathcal{Q}_0(A) \subseteq \mathcal{Q}_0(C)$, the same argument applies to any matrix in $\mathcal{Q}_0(A)$, so $\mathcal{Q}_0(A) \subseteq P_{0,n}$. $\square$

Example 7. There is no immediate converse to Corollary 1: for example, consider the matrix

$$A = \begin{pmatrix} 1 & 0 & 0 \\ 1 & 1 & 0 \\ 1 & 1 & 1 \end{pmatrix} .$$

By inspection, $\mathcal{Q}_0(A) \subset \mathcal{P}_{0,3}$, even though $A \notin \mathrm{CSD}$ (and hence $A + I \notin \mathrm{CSD}$) as a consequence of the submatrix

$$A(\{2,3\}|\{1,2\}) = \begin{pmatrix} 1 & 1 \\ 1 & 1 \end{pmatrix}$$

which is neither SNS nor SS.

Graph Theoretic Characterisation of CSD Matrices. CSD matrices have an elegant and simple graph-theoretic characterisation. Given any $n \times m$ matrix A we define a signed bipartite graph G_A as follows. G_A has n S-vertices associated with the rows of A, and m R-vertices associated with the columns of A. An edge exists between S-vertex i and R-vertex j if and only if $A_{ij} \neq 0$. The edge takes the sign of A_{ij}.

Let C be any cycle in G_A defined as a set of edges. The sign of C is

$$\mathrm{sign}(C) = \prod_{e \in C} \mathrm{sign}(e) .$$

As G_A is bipartite, $|C|$ is even, and one can define:

$$P(C) = (-1)^{|C|/2} \mathrm{sign}(C).$$

C will be termed an **o-cycle** if $P(C) = -1$. An example illustrating the definitions is shown in Figure 1.

Theorem 2. *Consider any matrix A and the associated graph G_A. The following two statements are equivalent:*

1. $A \in \mathrm{CSD}$;
2. All cycles in G_A are o-cycles.

Proof. This is proved in [11]. □

By Theorems 1 and 2, all matrices in the product $\mathcal{Q}_0(A)\mathcal{Q}_0(A^T)$ are P_0-matrices if and only if all cycles in G_A are o-cycles. Quite naturally, the same holds for all matrices in the product $\mathcal{Q}_0(A^T)\mathcal{Q}_0(A)$.

Example 8. Let A be any matrix of the form depicted in Figure 1. Taking the product of a matrix in $\mathcal{Q}_0(A^T)$ with one in $\mathcal{Q}_0(A)$ gives a matrix of the form

$$\begin{pmatrix} a_4 b_2 + a_1 b_1 & a_5 b_2 - a_2 b_1 \\ a_4 b_5 - a_1 b_3 & a_5 b_5 + a_3 b_4 + a_2 b_3 \end{pmatrix},$$

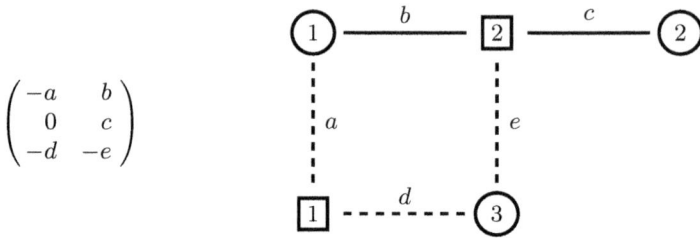

Fig. 1. A matrix and the corresponding signed bipartite graph. It is assumed that $a, b, c, d, e > 0$. Vertices corresponding to rows are depicted as circles and labelled with the row number; vertices corresponding to columns are depicted as squares and labelled with the column number. Negative edges are shown as dashed lines, while positive edges are shown as bold lines. Labels on the edges show the correspondence with entries in the matrix. The only cycle involves edges a, b, e, d. This has sign $(-1)^3 = -1$ and so the value of P for this cycle is $(-1^{4/2})(-1) = -1$. Thus this cycle is an o-cycle.

(where a_i, b_i are arbitrary nonnegative numbers.) By the results above any such matrix is necessarily a P_0 matrix. In fact, since 2×2 P matrices are positive stable, it lies in the closure of the positive stable matrices. Of course, both of these facts can be checked directly.

Theorem 2 gives a sufficient condition guaranteeing that an entire qualitative class consists of P_0-matrices:

Corollary 2. *Consider any square matrix A with nonnegative diagonal elements. If all cycles in G_{A+I} are o-cycles then $Q_0(A) \subset P_0$.*

Proof. Applying Theorem 2, all cycles in G_{A+I} are o-cycles, if and only if $A+I \in$ CSD. If $A + I \in$ CSD, then $C = Q_0(A + I) \subset$ CSD, and since $Q_0(A) + I \subseteq Q_0(A + I)$, $Q_0(A) + I \subset$ CSD. By Corollary 1, $Q_0(A) \subset P_0$. □

Remark 3. Corollary 2 has no immediate converse. However various conditions on the "interaction graph" of A, or a graph called the "directed SR graph" are equivalent to the statement $Q_0(A) \subset P_0$ [6,16]. It is interesting in general to ask when all matrices in a qualitative class have some property. For example, the question of when a qualitative class consists of matrices which are Hurwitz is posed in [12].

3.4   Compatibility of a Matrix with its Qualitative Class

A special case which arises frequently is when one of the factors in a product is constant, namely when, given functions $\mathcal{A} : X \to \mathbb{R}^{n \times m}$ and $\mathcal{B} : X \to \mathbb{R}^{n \times m}$, the function $\mathcal{A}$ is constant, i.e., $\mathcal{A}(x) = A$ for all $x \in X$. In particular we examine the situation where $\mathcal{B}(X) \subseteq Q_0(A^T)$, and are led to the question: "What are necessary and sufficient conditions on a matrix A to guarantee that all matrices in $Q_0(A)$ are compatible with A?"

Example 9. To see that compatibility of $\{A\}$ and $\mathcal{Q}_0(A)$ does not imply that every two matrices in $\mathcal{Q}_0(A)$ are compatible, consider the matrix

$$A = \begin{pmatrix} 1 & 1 \\ 1 & 1 \end{pmatrix}.$$

It is easy to check that A is compatible with each $B \in \mathcal{Q}_0(A)$, whereas

$$\begin{vmatrix} 1 & 2 \\ 2 & 1 \end{vmatrix} \begin{vmatrix} 2 & 1 \\ 1 & 2 \end{vmatrix} < 0,$$

and thus there are two matrices in $\mathcal{Q}(A)$ which fail to be compatible. Thus matrices which are compatible with their entire qualitative class form a larger set than the CSD matrices.

To motivate the discussion, consider functions $F(x) = Av(x)$, where A is a constant $n \times m$ matrix, $x \in X \subseteq \mathbb{R}^n$, $v : \mathbb{R}^n \to \mathbb{R}^m$ is a C^1 function, and v satisfies the condition $Dv(x) \in \mathcal{Q}_0(A^T)$ for each $x \in X$ (here $Dv(x)$ is the Jacobian matrix of v evaluated at x). Such functions arise in models from chemistry [7]. From arguments above, $DF(x) = A\,Dv(x) \in \mathcal{P}_0$ for each x and all allowed functions v if and only if all matrices in $\mathcal{Q}_0(A)$ are compatible with A.

Definition 6. *A matrix A will be termed **strongly sign determined (SSD)** if all square submatrices of A are either SNS or singular. $A \in$ SSD will mean that A is an SSD matrix, and $A \in \mathrm{SSD}_{n \times m}$ will mean that A is an $n \times m$ SSD matrix.*

The following two results show that the condition $A \in$ SSD is both necessary and sufficient to ensure that all matrices in $\mathcal{Q}_0(A)$ are compatible with A.

Lemma 10. *Consider an $n \times m$ matrix A and some $B \in \mathcal{Q}_0(A)$. If $A \in$ SSD, then (i) A and B are compatible; (ii) $AB^T \in \mathcal{P}_{0,n}$.*

Proof. Consider any $\alpha \subseteq \mathcal{I}_n$, $\beta \subseteq \mathcal{I}_m$ with $|\alpha| = |\beta|$. Since A is SSD, either $A[\alpha|\beta] = 0$ or $A(\alpha|\beta)$ is SNS. In the latter case, by Part 3 of Lemma 7 $\mathcal{Q}_0(A(\alpha|\beta)) \subseteq (\mathrm{SNS}_{|\alpha|} \cup \mathrm{SS}_{|\alpha|})$, so either $B[\alpha|\beta] = 0$, or $A[\alpha|\beta]B[\alpha|\beta] > 0$. In all cases $A[\alpha|\beta]B[\alpha|\beta] \geq 0$. Part (ii) now follows from Lemma 2. $\qquad \square$

Theorem 3. *Consider an $n \times m$ matrix A. Then every matrix $B \in \mathcal{Q}_0(A)$ is compatible with A if and only if A is SSD. Consequently $AB^T \in \mathcal{P}_{0,n}$ for every matrix $B \in \mathcal{Q}_0(A)$ if and only if A is SSD.*

Proof. By Lemma 10, if A is SSD, then each matrix in $\mathcal{Q}_0(A)$ is compatible with A, and $AB^T \in \mathcal{P}_{0,n}$. Conversely if A fails to be SSD, then there are sets $\alpha \subseteq \mathcal{I}_n$, $\beta \subseteq \mathcal{I}_m$ with $|\alpha| = |\beta|$, such that $A(\alpha|\beta)$ is neither SNS nor singular. By Lemma 6, there is some $B \in \mathcal{Q}(A)$ such that $A[\alpha|\beta]B[\beta|\alpha] < 0$. Thus A and B^T fail to be compatible. That we can choose B such that AB^T fails to be a P_0-matrix now follows from Lemma 4, since $\mathcal{Q}(A)$ is a sign-class and $\{A\}\mathcal{Q}(A^T)$ is an independent product. $\qquad \square$

Remark 4. Unlike the CSD matrices (Theorem 2), there is no known necessary and sufficient graph-theoretic characterisation of SSD matrices. However, a graph-theoretic condition, first presented in [17], was shown in [11] to be a sufficient condition for a matrix to be SSD. The construction is somewhat more involved than that for CSD matrices, requiring the introduction of edge-labels, and so is omitted here. It can be regarded as a special case of a very general sufficient graph-theoretic condition for a product of two matrices to be a P_0-matrix [8].

Example 10. An $n \times m$ matrix in which every entry is the same is (trivially) SSD, since every square submatrix is singular. However if $n, m > 1$, and the entries are nonzero, then it is not CSD. This was illustrated in Example 9.

Remark 5. Parts 4 and 5 of Lemma 7 imply that both $\mathrm{CSD}_{n \times m}$ and $\mathrm{SSD}_{n \times m}$ are closed. Moreover, $\mathrm{CSD}_{n \times m}$ (but not $\mathrm{SSD}_{n \times m}$) consists of entire qualitative classes. One interesting consequence is that some chemical reaction network structures do not permit multiple equilibria, regardless both of the kinetics *and of the stoichiometries.* The frequency with which CSD/SSD matrices occur among $n \times m$ matrices whose entries are small integers is an interesting question worthy of some exploration.

3.5   Matrices with Diagonal Factors

We now consider matrix products of the form AD_mBD_n, where A is an $n \times m$ matrix, B is an $m \times n$ matrix, and $D_m \in \mathcal{D}_m$, $D_n \in \mathcal{D}_n$. There are a variety of situations in which such products arise, including chemical models involving mass-action and generalised mass-action kinetics. To motivate the discussion, consider the function $F(x) = Av(x)$, where A is a constant $n \times m$ matrix, $x \in X \subseteq \mathbb{R}^n$, $v : \mathbb{R}^n \to \mathbb{R}^m$, and components of v take the form:

$$v_j(x) = \phi_j \left(\sum_{k=1}^n B_{jk} g_k(x_k) \right). \tag{1}$$

Here $\phi_j(\cdot)$ and $g_k(\cdot)$ are C^1 functions satisfying $\phi_j'(\cdot) > 0$ and $g_k'(\cdot) > 0$, and B_{jk} are constants.

Example 11. Defining X to be the interior of the nonnegative orthant in $\mathbb{R}^n$, and choosing $g_k(x) = \ln x$ and $\phi_j(y) = K_j e^y$, gives

$$v_j(x) = K_j \prod_{k=1,\ldots,n} x_k^{B_{jk}}$$

which is the functional form for the rates of reactions assuming "generalised mass-action kinetics". The particular choice $B_{jk} = \max\{-S_{kj}, 0\}$ gives mass-action kinetics.

Differentiating (1) gives

$$\frac{\partial v_j}{\partial x_i} = \phi'_j \left(\sum_{k=1}^{n} B_{jk} g_k(x_k) \right) B_{ji} g'_i(x_i) \,.$$

So $Dv(x) = D_1(x)BD_2(x)$, where D_1 and D_2 are positive diagonal matrices with entries $(D_1)_{jj} = \phi'_j \left(\sum_{k=1}^{n} B_{jk} g_k(x_k) \right)$ and $(D_2)_{ii} = g'_i(x_i)$. Thus $DF(x) = AD_1(x)BD_2(x)$.

Lemma 11. *Two $n \times m$ matrices A and B are compatible if and only if A and $D_n B D_m$ are compatible for every $D_n \in \mathcal{D}_n$ and $D_m \in \mathcal{D}_m$.*

Proof. Since only principal minors of a diagonal matrix can be nonzero, for any $\alpha \subseteq \mathcal{I}_n$ and $\beta \subseteq \mathcal{I}_m$,

$$(D_n B D_m)[\alpha|\beta] = D_n[\alpha] B[\alpha|\beta] D_m[\beta] \,.$$

As $D_n[\alpha]$ and $D_m[\beta]$ are positive, $(D_n B D_m)[\alpha|\beta]$ has the same sign as $B[\alpha|\beta]$. $\square$

Lemma 12. *Consider $n \times m$ matrices A and B. If A and B are compatible, then $AD_m B^T D_n \in \mathcal{P}_{0,n}$ for any $D_m \in \mathcal{D}_m$ and any $D_n \in \mathcal{D}_n$.*

Proof. By Lemma 11, since B is compatible with A, $(D_n B D_m)$ is compatible with A. By Lemma 2, $AD_m B^T D_n = A(D_n B D_m)^T \in \mathcal{P}_{0,n}$. $\square$

Remark 6. Given a matrix A, define A_-, the **negative part of** A, via $(A_-)_{ij} = \min\{A_{ij}, 0\}$. Matrices compatible with their negative parts were termed "weakly sign determined" (WSD) in [7]. By Lemma 10, matrices which are SSD are WSD, but the converse need not be true. Theorem 4.1 in [7] can be seen as an application of Lemma 12.

Example 12. The matrices

$$A = \begin{pmatrix} 1 & -2 \\ -1 & 1 \end{pmatrix} \quad \text{and} \quad A_- = \begin{pmatrix} 0 & -2 \\ -1 & 0 \end{pmatrix}$$

are compatible, even though A fails to be SSD. Consequently, by Lemma 12, $AD_1(A_-)^T D_2 \in \mathcal{P}_{0,n}$ for any $D_1, D_2 \in \mathcal{D}_2$.

Lemma 13. *Consider $n \times m$ matrices A and B, which are not compatible. Then given any fixed $D_n \in \mathcal{D}_n$, there exists $D_m \in \mathcal{D}_m$ such that $AD_m B^T D_n \notin \mathcal{P}_{0,n}$.*

Proof. By Lemma 11, A and $D_n B D_m$ are not compatible for any positive diagonal matrices D_n, D_m. Further, $\mathcal{C} = \{D_m B^T D_n \mid D_m \in \mathcal{D}_m\}$ is a left sign-class. Moreover, $\{A\}\mathcal{C}$ is an independent product. So, by Lemma 3, there exists a D_m such that $AD_m B^T D_n \notin \mathcal{P}_{0,n}$. $\square$

Remark 7. Theorem 4.3 in [7] is an application of this result.

A special case is where we have $B = A^T$. In this case, the results of Lemma 12 can be strengthened:

Theorem 4. *Let* $D_m \in \mathcal{D}_m$, $D_n, \tilde{D}_n \in \mathcal{D}_n$, *and* A *be an* $n \times m$ *matrix. Then (i)* $AD_m A^T D_n + \tilde{D}_n$ *is positive stable; (ii)* $AD_m A^T D_n$ *lies in the closure of the positive stable matrices.*

Proof. Since A is (trivially) compatible with itself, we immediately have from Lemma 12 that $AD_m A^T D_n \in \mathcal{P}_{0,n}$, and so $AD_m A^T D_n + \tilde{D}_n$ is a P-matrix. Secondly, $J = AD_m A^T D_n + \tilde{D}_n$ is "sign-symmetric", i.e., given any $\alpha \subseteq \mathcal{I}_n$, $\gamma \subseteq \mathcal{I}_m$ with $|\alpha| = |\gamma|$, $J[\alpha|\gamma]J[\gamma|\alpha] \geq 0$. To see this, rewrite $J = (AD_m A^T + \tilde{D}_n D_n^{-1})D_n$, and note that $C \equiv AD_m A^T + \tilde{D}_n D_n^{-1}$ is symmetric. Applying the Cauchy-Binet formula, we have

$$J[\alpha|\gamma]J[\gamma|\alpha] = C[\alpha|\gamma]C[\gamma|\alpha]D_n[\gamma]D_n[\alpha] = (C[\alpha|\gamma])^2 D_n[\gamma]D_n[\alpha] \geq 0\,.$$

Since sign-symmetric P-matrices are positive stable (Theorem 1.1 in [10]), it follows that $AD_m A^T D_n + \tilde{D}_n$ is positive stable. The second claim follows immediately because $\tilde{D}_n$ can have arbitrarily small diagonal entries. □

Remark 8. The example in Section 4.4 of [18] can be regarded as an application of this result. As usual in applications, it is the dual result which appears, and certain Jacobian matrices arising in biochemistry are shown to be structurally Hurwitz. In general, examples of systems in biology/chemistry which are structurally Hurwitz are few and far between. These systems provide an exception.

4   Discussion and Conclusions

A variety of sufficient and necessary conditions have been found for the product of matrices to lie in the closure of the P-matrices. In some cases, it has been shown that (positive) stability of the product follows. Underlying most of these results are the notion of compatibility, and application of the Cauchy-Binet formula. The results themselves have been shown to have application to questions of injectivity and stability in situations where Jacobian matrices have a natural product structure as, for example, occurs often in biology and chemistry. A number of examples have been presented to illustrate the basic concepts, and it has been remarked that several previously obtained results are special cases of those presented here.

References

1. Murty, K.G.: Linear Complementarity, Linear and Nonlinear Programming. Heldermann Verlag, Berlin (1988)
2. Hofbauer, J., Sigmund, K.: Evolutionary games and population dynamics. Cambridge University Press (1998)
3. Gale, D., Nikaido, H.: The Jacobian matrix and global univalence of mappings. Math. Ann. 159, 81–93 (1965)

4. Parthasarathy, T.: On global univalence theorems. Lecture Notes in Mathematics, vol. 977. Springer, Heidelberg (1983)
5. Nikaido, H.: Convex structures and economic theory. Academic Press (1968)
6. Soulé, C.: Graphic requirements for multistationarity. Complexus 1, 123–133 (2003)
7. Banaji, M., Donnell, P., Baigent, S.: P matrix properties, injectivity and stability in chemical reaction systems. SIAM J. Appl. Math. 67(6), 1523–1547 (2007)
8. Banaji, M., Craciun, G.: Graph-theoretic approaches to injectivity and multiple equilibria in systems of interacting elements. Commun. Math. Sci. 7(4), 867–900 (2009)
9. Kellogg, R.B.: On complex eigenvalues of M and P matrices. Numer. Math. 19, 70–175 (1972)
10. Hershkowitz, D., Keller, N.: Positivity of principal minors, sign symmetry and stability. Linear Algebra Appl. 364, 105–124 (2003)
11. Banaji, M., Craciun, G.: Graph-theoretic criteria for injectivity and unique equilibria in general chemical reaction systems. Adv. in Appl. Math. 44, 168–184 (2010)
12. Maybee, J., Quirk, J.: Qualitative problems in matrix theory. SIAM Rev. 11(1), 30–51 (1969)
13. Brualdi, R.A., Shader, B.L.: Matrices of sign-solvable linear systems. Cambridge tracts in mathematics, vol. 116. Cambridge University Press (1995)
14. Gantmacher, F.R.: The theory of matrices. Chelsea (1959)
15. Banaji, M., Rutherford, C.: P-matrices and signed digraphs. Discrete Math. 311(4), 295–301 (2011)
16. Banaji, M.: Graph-theoretic conditions for injectivity of functions on rectangular domains. J. Math. Anal. Appl. 370, 302–311 (2010)
17. Craciun, G., Feinberg, M.: Multiple equilibria in complex chemical reaction networks: II. The species-reaction graph. SIAM J. Appl. Math. 66(4), 1321–1338 (2006)
18. Donnell, P., Banaji, M., Baigent, S.: Stability in generic mitochondrial models. J. Math. Chem. 46(2), 322–339 (2009)

A Formal Model for Databases in DNA

Joris J.M. Gillis* and Jan Van den Bussche

Hasselt University and Transnational University of Limburg,
Agoralaan Gebouw D, 3590 Diepenbeek, Belgium

Abstract. Our goal is to better understand, at a theoretical level, the
database aspects of DNA computing. Thereto, we introduce a formally
defined data model of so-called *sticker DNA complexes*, suitable for the
representation and manipulation of structured data in DNA. We also
define DNAQL, a restricted programming language over sticker DNA
complexes. DNAQL stands to general DNA computing as the standard
relational algebra for relational databases stands to general-purpose conventional computing. The number of operations performed during the
execution of a DNAQL program, on any input, is only polynomial in
the *dimension* of the data, i.e., the number of bits needed to represent
a single data entry. Moreover, each operation can be implemented in
DNA using a constant number of laboratory steps. We prove that the
relational algebra can be simulated in DNAQL.

Keywords: DNA Computing, Formal Model, Relational Algebra.

1   Introduction

In DNA computing [16,3], data are represented using synthetic DNA molecules
in vitro. Operations on data are performed by biotechnological manipulations
of DNA that are based on DNA self-assembly (Watson–Crick base pairing) or
on explicit effects upon DNA by specific enzymes. In the original approach to
DNA computing, which we could call the Adleman model [2,5,18], one uses a
more or less standard repertoire of operations on DNA, where each operation
corresponds to a fixed number of steps in the laboratory. (These steps could be
performed by a human or by a robot.)

In more recent years, research in DNA computing is largely focusing on the
goal to let an entire computation happen by self-assembly alone, without (or with
minimal) outside intervention, e.g., [23,22,11]. Whereas the pure self-assembly
model is very attractive, it is harder to achieve in practice, and indeed this is
the subject of a lot of current research in the area of DNA nanotechnology.

Meanwhile, the original Adleman model deserves further study, and in this paper we have a renewed look at the Adleman model, specifically from the perspective of databases. Indeed, DNA computing is very attractive from the database
perspective: the nanoscale and robustness of DNA molecules are promising from

---

* Ph.D. fellowship of the Research Foundation - Flanders (FWO)

K. Horimoto, M. Nakatsu, and N. Popov (Eds.): ANB 2011, LNCS 6479, pp. 18–37, 2012.

a data storage point of view, and the highly parallel operations of DNA computing correspond well with the bulk data processing nature typical of database query processing [1]. Most earlier theoretical work on the possibilities of DNA computing has focused either on the ability to mimick classical models of computation such as finite automata or Turing machines, or on the relationship with parallel computation, but this always from a general-purpose computing perspective.

In contrast, in database theory, one considers restricted models of computation, limited in computational power but still with sufficient expressiveness for structured database manipulation. The classical model is the relational algebra for relational databases [1]. This algebra consists of six operations on relations (database tables): union; difference; cartesian product; selection; projection; and renaming. These operations can be composed to form expressions. These express database queries, and the relational algebra can express precisely those database queries that can be defined in first-order logic, thus providing a well-delineated restriction in computational power.

The benefit of restricted computational models is that they facilitate the identification of optimisation strategies for more efficient processing; hence there exists a large body of techniques for database query processing, e.g., [17]. From the point of view of theoretical science, an added benefit of a restricted computational model is that it allows us to study and attempt to characterise the precise computational abilities of the computational systems that are being modeled (such as relational database systems).

Motivated by the above considerations, in this paper, we want to propose a solution to the following equation:

$$\frac{\text{relational databases and relational algebra}}{\text{general-purpose conventional computing}} = \frac{?}{\text{DNA computing (Adleman model)}}$$

We define a formal data model of *sticker complexes*, which represent complexes of DNA molecules. Our complexes are general enough to serve as data structures for structured data such as found in relational databases. At the same time, however, sticker complexes are restricted so that we avoid the complications connected to the difficult secundary structure prediction problem of general DNA complexes [14]. Indeed, our main contribution consists in formally defining a well-behaved family of DNA-complex data structures, with an accompanying set of operations on these data structure that preserve the well-behavedness restrictions. We fit the operations into a first-order query language, called DNAQL, with a formal operational semantics. We thus propose the sticker complex data model, together with DNAQL, as the DNA computing analogues of the relational database model and the accompanying relational algebra. Restrictive as sticker complexes and DNAQL may be, we prove that they can still simulate the relational data model and the relational algebra. At the same time, we stress that our new DNA database model should also be appreciated in its own right as a restricted model of DNA computing specialised to database manipulation.

This paper is organised as follows. Section 2 discusses related work. Section 3 defines the data model. Section 4 introduces important operations on sticker complexes. Section 5 discusses the representation of structured data using complexes. Section 6 discusses the implementation in DNA of the operations. Section 7 defines the query language DNAQL. Section 8 presentes the simulation of the relational algebra in DNAQL. We conclude in Section 9.

2   Related Work

Our work can be seen as a followup of Reif's original work [18] on relating DNA computing with conventional parallel computing. Indeed, Reif also formalized DNA complexes and considered similar operations. Our model specializes Reif's model to a database model. For example, it is well known [1,12] that the data complexity of the relational algebra (first-order logic) belongs to the parallel circuit complexity class AC_0, denoting constant-depth, polynomial-size circuits with unbounded fan-in. Likewise, DNAQL programs execute a number of operations on complexes that are largely independent of the data size, except for a polynomial dependence on the number of bits needed to represent a single data entry, a number we call the *dimension* of the data. Moreover, as usual for DNA computing, each operation works in parallel on the different DNA strands present in a complex, and each operation can be implemented in real DNA in a constant number of laboratory steps.

Our work also clearly fits in a recent trend in DNA computing to identify specialised computational models within the general framework of DNA computing. This trend is nicely exemplified by the work by Cardelli [6] and Majumder and Reif [15], where the specialised computational model is that of process algebras; in our work, it is that of databases.

While our work is not the first to relate the relational algebra with DNA computing, we are the first to do it formally and in detail. An abbreviated account of achieving relational algebra operations through DNA manipulation was given recently by Yamamoto et al. [26], but unfortunately that paper is too sketchy to allow any comparison with our approach. In contrast, our own methods are fully formalised, and importantly, our work identifies restrictions on DNA computing within which relational algebra simulation remains possible. More influential to our work is the older work by Arita et al. [4] demonstrating how one can accomplish concatenation and rotation of DNA strands. Such manipulations, which involve circular DNA, are crucial in our model, and indeed were already crucial to Reif [18].

Finally, we mention the earlier works on DNA memories [19,7], which, while having a database flavor, are primarily about supporting *searching* in sets of DNA strands and largely ignore the more complex operations of the relational algebra such as difference, projection, cartesian product, and renaming.

3   The Sticker-Complex Data Model

In this section we formally define a family of data structures which we call *sticker complexes*. They are an abstraction of complexes of DNA strands. Reif

[18] already defined a similar data structure, but our definition introduces several limitations so as to avoid unrealistic or otherwise complicated and unmanageable secundary structures. The adjective 'sticker' points to our restriction of hybridization to short primers (which we call "negative" strands) for the recognition and splicing of the strands carrying the actual data (called the "positive" strands).

Basically, we assume the following disjoint, finite alphabets: Λ of *atomic value symbols*; Ω of *attribute names*; and $\Theta = \{\#_1, \#_2, \#_3, \#_4, \#_5, \#_6, \#_7, \#_8, \#_9\}$ of *tags*. The union of these three alphabets is denoted by Σ and called the *positive alphabet*.

Furthermore, we use a *negative alphabet*, denoted $\overline{\Sigma}$, disjoint from Σ, defined as $\overline{\Sigma} = \{\bar{a} \mid a \in \Sigma\}$. Thus there is a bijection between Σ and $\overline{\Sigma}$, which is called *complementarity* and is denoted by overlining a symbol; we set $\bar{\bar{a}} = a$.

We will first define *pre-complexes* that contain the overall structure of sticker complexes. Sticker complexes will then be defined as pre-complexes satisfying various restrictions. A pre-complex is a finite, edge-labeled directed graph where the edges represent bases in strands, and the nodes represent the endpoints between the bases in a strand. Moreover, a pre-complex is equipped with a matching, representing base pairing, and two predicates. One predicate indicates which bases are "immobilized", i.e., do not float freely and can be separated from solution in a controlled manner; the other predicate indicates which bases are "blocked", i.e., cannot participate in base pairing. Formally, a pre-complex is a 6-tuple $(V, E, \lambda, \mu, immob, blocked)$ such that:

1. V is a finite set of nodes,
2. $E \subseteq V \times V$ is a finite set of directed edges without self-loops,
3. $\lambda : E \to \Sigma \cup \overline{\Sigma}$ is a total function labeling the edges,
4. $\mu \subseteq [E]^2 = \{\{e, e'\} \mid e, e' \in E \text{ and } e \neq e'\}$ is a partial matching on the edges, i.e., each edge occurs in at most one pair in μ,
5. $immob \subseteq E$,
6. $blocked \subseteq E$.

Let C be a pre-complex as above. We introduce the notion of "strand" and "component" of C as follows. A *strand* of C is simply a connected component of the directed graph (V, E). Furthermore, we say two strands s and s' are *bonded* if there exists some edge e in s and some edge e' in s' with $\{e, e'\} \in \mu$. When two strands are connected (possibly indirectly) by this bonding relation, we say they belong to the same component. Thus, a *component* of a pre-complex is a substructure formed by a maximal set of strands connected by the bonding relation.

A *sticker complex* now is a pre-complex satisfying the following restrictions:

1. There are no isolated nodes, i.e., each node occurs in at least one edge.
2. Each node has at most one incoming and at most one outgoing edge. Thus, each strand has the form of a chain or a cycle.
3. The labels on a chain are "homogeneous", in the sense that either all edges are labeled with positive symbols or all edges are labeled with negative

symbols. Naturally, a strand with positive (negative) symbols is called a positive (negative) strand.

4. Negative strands are severely restricted: specifically, every negative strand must be a chain of one or two edges.
5. Matchings by μ can only occur between complementarly labeled edges.
6. An edge can be immobilized only if it is the sole edge of a negative strand.
7. Edges in *blocked* do not occur in μ.
8. Each component can contain at most one immobilized edge.

Henceforth, for simplicity, we will refer to sticker complexes simply as "complexes".

We remark that the predicate *blocked* and the matching μ serve to abstract two different features of double-strandedness. The matching μ is used to make explicit where the stickers (short negative strands) pair with the positive strands. The predicate blocked represents longer stretches of double strands. As in the work by Rozenberg and Spaink [20], blocking is used to restrict the places where hybridization can still occur.

We also remark that it is not necessary to require that edges matched by μ run in opposite directions (in accordance with the opposite 5'–3' and 3'–5' directions of double-stranded DNA). This is because stickers of length one can trivially be placed in the desired direction, and stickers of length two can always fold so as to be again in the desired direction. The latter is illustrated in Figure 1.

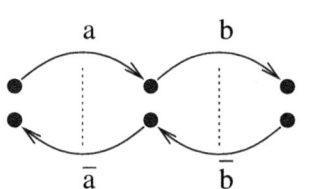
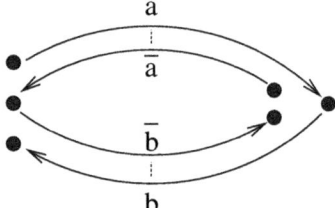

Fig. 1. On the left, a complex with two strands spelling the words ab and $\bar{b}\bar{a}$ and the expected complementary base pairing. On the right, a complex with two strands spelling the words ab and $\bar{a}\bar{b}$ and a "folded" base pairing. Dotted lines denote edges matched by μ.

Redundancy in complexes. In practice, a test tube will contain many duplicate strands, and indeed this multiplicity is typically crucial for DNA computing to work. Accordingly, in our model, each component of a complex stands for possibly multiple occurrences. (This important issue is not addressed in Reif's formalisation of complexes [18].) In order to formalize this, we define the notions of subsumption, equivalence, redundant extension, and minimality.

A complex C' is said to *subsume* a complex C if for each component D of C, there exists an isomorphic component D' in C'. Two complexes C and C' are said to be *equivalent* if they subsume each other. When C' is equivalent to C and an extension of C, we call C' a *redundant extension* of C.

A component D of a complex C is called redundant if some other component of C is isomorphic to D. Note that removing a redundant component from C yields a complex that is still equivalent to C. A complex that has no redundant components is called *minimal*. Naturally, each complex C has a unique (up to isomorphism) minimal complex C' that is equivalent to C; we call C' the *minimization* of C.

4   Operations on Complexes

In this section, we formally define a set of operations on complexes that are rather standard in the DNA computing literature, except perhaps the difference. But what is interesting, however, is that we have defined sticker complexes in such a way that each operation always result in a sticker complex when applied to sticker complexes. Moreover, the difference operation imposes additional restrictions on its input so as to guarantee effective implementability in real DNA (discussed in Section 6).

As a general proviso, in the following definitions, a final minimization step should always be applied to the result so as to obtain a mathematically deterministic operation. In the following definitions we keep this implicit so as not to clutter up the presentation. Also, it is understood that the result of each operation is defined up to isomorphism.

Union. Let $C_1 = (V_1, E_1, \lambda_1, \mu_1, immob_1, blocked_1)$ and $C_2 = (V_2, E_2, \lambda_2, \mu_2, immob_2, blocked_2)$ be two complexes. W.l.o.g. we assume that V_1 and V_2 are disjoint. Then the union $C_1 \cup C_2$ equals $(V_1 \cup V_2, E_1 \cup E_2, \lambda_1 \cup \lambda_2, \mu_1 \cup \mu_2, immob_1 \cup immob_2, blocked_1 \cup blocked_2)$.

Difference. Let C_1 and C_2 be two complexes that satisfy the following conditions:

1. $\mu_1 = immob_1 = blocked_1 = \emptyset = \mu_2 = immob_2 = blocked_2$, i.e., all components in C_1 and C_2 are single strands.
2. All strands of C_1 and C_2 are positive, noncircular, and all have the same length.
3. Each strand of C_2 ends with $\#_4$ and does not contain $\#_5$.

Then the difference $C_1 - C_2$ equals the union of all strands in C_1 that do not have an isomorphic copy in C_2. If C_1 and C_2 do not satisfy the above conditions then $C_1 - C_2$ is undefined.

Hybridize. Let $C = (V, E, \lambda, \mu, immob, blocked)$ and $C' = (V', E', \lambda', \mu', immob', blocked')$ be two complexes. We say that C' is a *hybridization extension* of C if $V = V'$, $E = E'$, $\lambda = \lambda'$, $immob = immob'$, $blocked = blocked'$ and μ' is an extension of μ. Beware that a hybridization extension must satisfy all conditions from the definition of sticker complex. A complex C' is said to have *maximal matching* if the only hybridization extension of C' is C' itself.

The notion of hybridization extension is not sufficient, however, since we want to allow duplicate copies of components in C to participate in hybridization. (This important issue is glossed over in Reif's formalisation [18].) To formalize this behavior, let us call C' (with matching μ') a *multiplying hybridization extension (MHE)* of C if C' is a hybridization extension, with maximal matching, of some redundant extension C'' of C. Moreover, we call a component D of an MHE *unfinished* if there exist another MHE in which D occurs bonded within a larger component. We then call an MHE *saturated* if it has no unfinished components. This is illustrated in Figure 2.

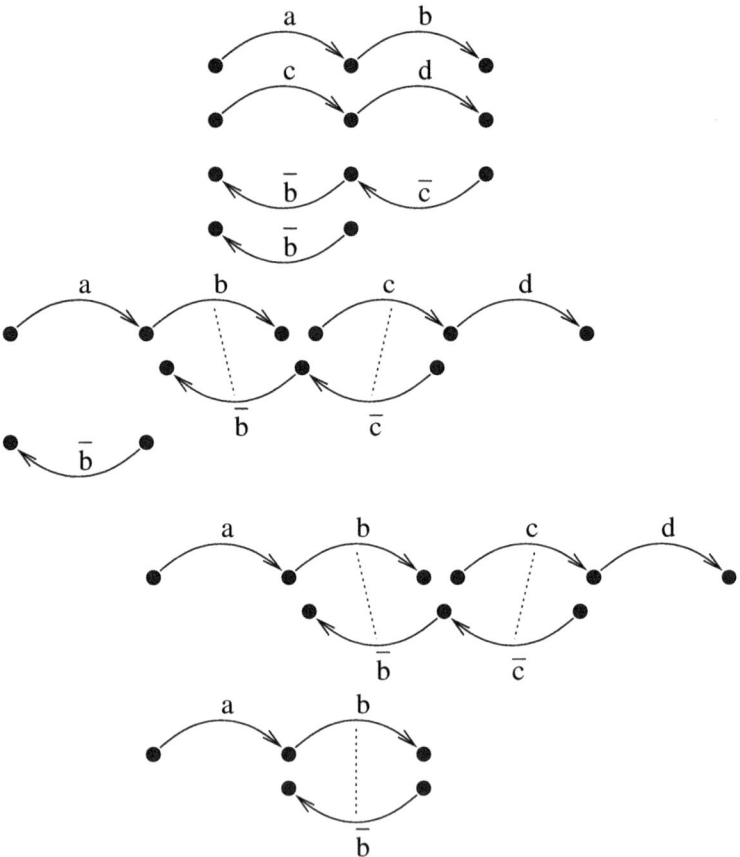

Fig. 2. Left: a complex C; top right: a hybridization extension of C with maximal matching, but not saturated in view of the MHE of C shown bottom right; that MHE is saturated. Dotted lines denote edges matched by μ.

Finally we say that C has *recursion-free hybridization* if there exists only a finite number of saturated hybridization extensions of C.

On the other hand, we do not want hybridization to go off into an uncontrolled chain reaction. Indeed, our very goal in this paper is to explore a "first-order" or "recursion-free" version of DNA computing, in line with the first-order nature of the relational algebra [1]. Thus we want to stay away from recursive self-assembly DNA computations. Formally, we want to rule out the situations where there are infinitely many possible non-equivalent MHE's. Such situations are very well possible. Consider, for a simple example, the complex C consisting of two non-circular strands spelling out the words ab and $\bar{a}\bar{b}$. Taking n copies of ab and n copies of $\bar{a}\bar{b}$, we can form arbitrary long non-equivalent MHE's of C. An illustration for $n = 3$ is given in Figure 3.

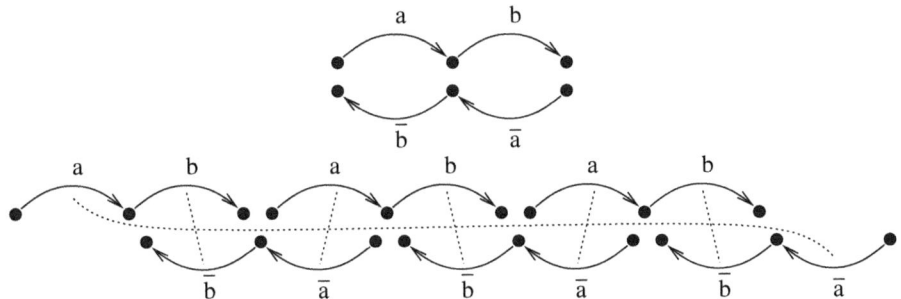

Fig. 3. A complex (top) and one of its MHE's (bottom). Dotted lines denote edges matched by μ. Note that the MHE forms a ring structure.

Formally, we say that C has *recursion-free hybridization* if their are only finitely many saturated MHE's of C. If this is the case, we define `hybridize`(C) to equal the disjoint union of all saturated MHE's of C. If C does not have recursion-free hybridization, we consider `hybridize`(C) to be undefined. For example, it can be verified that the complex from Figure 2 has recursion-free hybridization.

Ligate. The ligate operator concatenates strands that are held together by a sticker. Formally, define a *gap* as a set of four edges $\{e_1, e_2, e_3, e_4\}$ such that $\{e_1, e_4\} \in \mu$; $\{e_2, e_3\} \in \mu$; e_1 and e_2 (in that order) are consecutive edges on a negative strand; e_3 is the last edge on its (positive) strand; and e_4 is the first edge on its (positive) strand. By *filling a gap* we mean modifying the complex so that the endnode of e_3 and the startnode of e_4 are identified. We now define `ligate`(C) as the complex obtained from C by filling all gaps.

Flush. Quite simply `flush`(C) equals the complex obtained from C by removing all components that do not contain an immobilized edge.

Split. Consider a node u in some complex C. By *splitting C at u*, we mean the following.

- If u has an incoming (outgoing) edge, denote it by e_1 (e_2).
- If both e_1 and e_2 exist, then replace u by two nodes u_1 and u_2, letting e_1 arrive in u_1, and letting e_2 start in u_2.
- Furthermore, if there exists a node u' with incoming edge e_4 and outgoing edge e_3, such that $\{e_1, e_3\} \in \mu$ or $\{e_2, e_4\} \in \mu$, then u' is also split in an analogous manner.

Also, an edge is called *interacting* if it neither occurs in *blocked* nor in μ.

Now consider the set of triples shown in Table 1. Each triple is called a *split-point* and has the form (*label*, *interacting*, *place*). By splitting C at such a split-point, we mean splitting C at all startnodes (if *place* is 'before') or endnodes (otherwise) of edges labeled *label*, on condition that the edge is interacting (or noninteracting, depending on the boolean value *interacting*). The result is denoted by $\mathtt{split}(C, label)$.

Table 1. The allowed split points

Label	Interacting	Place
$\#_2$	*false*	before
$\#_3$	*false*	before
$\#_4$	*false*	after
$\#_6$	*true*	after
$\#_8$	*true*	before

Blocking. There are two blocking operations. Here we assume that C is "saturated" in the sense that C is equivalent to $\mathtt{hybridize}(C)$; if this condition is not satisfied then the blocking operations on C are considered to be undefined.

The simplest operation is $\mathtt{block}(C, \sigma)$, for any $\sigma \in \Sigma$, which equals the complex obtained from C by adding all edges labeled σ to *blocked*.

For the other operation, let again be $\sigma \in \Sigma$, and consider any contiguous substrand s in C. We call s a σ-*blocking range* if it satisfies three conditions. Firstly, all edges of the substrand are interacting (in the sense of the previous paragraph). Secondly, either the substrand contains the first edge of its strand, or the edge preceding the first edge of the substrand is blocked. Thirdly, the last edge of the substrand is labeled with σ. Now we define $\mathtt{blockfrom}(C, \sigma)$ to be the complex obtained from C by adding to *blocked* all edges appearing in some σ-blocking range.

Cleanup. The cleanup operator undoes matchings and blockings and removes all strands except for the longest positive strands. Here we assume the condition that every positive strand in C is at least three long, and has at least one interacting edge; if C does not satisfy this condition, $\mathtt{cleanup}(C)$ is not defined. Otherwise, $\mathtt{cleanup}(C)$ equals the union of all positive strands of C of maximal length; there are no matched and no blocked edges in $\mathtt{cleanup}(C)$.

5   Data Representation

When we want to represent structured data as sticker complexes, the symbols from the alphabet $\Sigma = \Lambda \cup \Omega \cup \Theta$ will be used in different ways. Attributes (Ω) will be used to indicate the structure of the data; tags (Θ) will be used as separators and auxiliary markers in data manipulation. Atomic value symbols (Λ) will be used to represent the actual data entries. However, since Λ is just a finite alphabet typically of small size, we will need to use strings (or vectors) of atomic value symbols to represent data entries, just like words of bits are used in conventional computing to represent data entries like characters or integers. In analogy to the word length of a conventional computer processor, in our approach we assume some *dimension* ℓ, a natural number, is known. Then every data entry is encoded by an ℓ-vector of atomic data symbols.

Formally, we say that a sticker complex C *has dimension* ℓ if every edge e labeled by some (positive) atomic value symbol is part of a sequence $(e_0, e_1, \ldots, e_\ell, e_{\ell+1})$ of $\ell + 2$ consecutive edges, where e_0 is labeled $\#_3$; each e_i for $i = 1, \ldots, \ell$ is labeled with a positive atomic value symbol; and $e_{\ell+1}$ is labeled $\#_4$. So, e is one of the e_i's with $i \in \{1, \ldots, \ell\}$. We call $(e_0, e_1, \ldots, e_\ell, e_{\ell+1})$ an ℓ-*vector* in C. A complex of dimension ℓ is also called an ℓ-*complex*.

We also introduce an additional blocking operator on ℓ-complexes. Let n be a natural number and let C be a complex satisfying the following conditions:

1. C is an ℓ-complex with $\ell \geq n$;
2. in every ℓ-vector in C, either all edges are blocked or no edge is blocked;
3. C is equivalent to `hybridize`(C).

Then `blockexcept`(C, n) equals the complex obtained from C by blocking, within each ℓ-vector $(e_0, e_1, \ldots, e_\ell, e_{\ell+1})$ that is not yet blocked, all edges except e_n. If (C, n) does not satisfy the conditions above, then `blockexcept`(C, n) is undefined.

6   Implementation in DNA

In this section, we argue that the abstract sticker complexes and the operations on them presented above can be implemented by real DNA complexes. Our discussion remains theoretical as we have not performed laboratory experiments. On the one hand, our main purpose is to make the abstract model plausible as a theoretical framework in which the possibilities and limitations of DNA computing as a database model; on the other hand, we use only rather standard biotechnological techniques.

Each component of an abstract complex is represented by a large surplus of duplicate copies in DNA. Each positive alphabet symbol from Σ is implemented by a strand of (single-stranded) DNA, such that the resulting set of DNA strands forms a set of DNA codewords [8,21,24]. If the DNA strand for symbol $a \in \Sigma$ is w, then the DNA strand for the complementary symbol $\bar{a}$, is, naturally, the Watson-Crick complementary strand to w. Then, matching of edges by μ in an abstract complex is implemented by base pairing in the DNA complex. We will see below how blocking is implemented. Immobilization is implemented as is standard in DNA computing by attachment to surfaces [13] or magnetic beads.

The union operation amounts to mixing two test tubes together.

The difference $C_1 - C_2$ of complexes can be implemented by a subtractive hybridization technique [10]. Let C_1 (C_2) be stored in test tube t_1 (t_2). Because all strands in t_2 end in $\#_4$, we can easily append $\#_5$ to them. Next we add to t_2 an abundance of immobilized short primers $\overline{\#_5}$. Using polymerase we obtain complements to all strands in t_2, still immobilized, so that it is now easy to separate them. It remains to use these complements to remove all strands from t_1 that occured in t_2. Since all strands have the same length, partial hybridization, leading to false removals, can be avoided by using a very precise melting temperature based on the precise length of the strands.

Hybridization happens naturally and is merely controlled by temperature. Still, we must argue that the result still satisfies the definition of sticker complex. The only peculiarity in this respect is the requirement that each component can contain at most immobilized edge. Since immobilized edges are implemented by strands affixed to surfaces, implying some minimal distance between such strands, it seems reasonable to assume that the large majority of hybridization reactions will occur among freely floating strands, or between freely floating and immobilized ones.

Recursion-free hybridization is very hard to control by nature. It will be the responsability of the algorithm designer to design DNAQL programs (see Section 7) that, on the intended inputs, will apply `hybridize` only to inputs that have recursion-free hybridization. Our simulation of the relational algebra in DNAQL (see Section 8) is well-defined in this sense.

Splitting is achieved as usual by restriction enzymes. A feature of our abstract model is that we require only five recognition sites (Table 1). Of course, these recognition sites will have to be integrated in the DNA codeword design.

Blocking is implemented by making strands double-stranded, so that they cannot be involved in later hybridizations. The ordinary `block` operation can be implemented by adding the appropriate primer which will anneal to the desired substrands thus blocking the corresponding edges. As in the Sanger sequencing method, however, the base at the 3' end of the primer is modified to its dideoxy-variant. In this way unwanted interaction with polymerase from possible later `blockfrom` operations is avoided. Indeed, `blockfrom` is implemented using polymerase.

For the `blockexcept` operation to work, we need to adapt the implementation of ℓ-vector strands $\#_3 v_1 \ldots v_\ell \#_4$, with $v_i \in \Lambda$ for $i = 1, \ldots, \ell$, by introducing additional markers ϕ_i, so that we get $\#_3 \phi_1 v_1 \ldots \phi_\ell v_\ell \#_4$. These ℓ additional markers must be part of the set of codewords. We can then implement `blockexcept(., n)` by the composition `block(., #_3)`; `blockfrom(., ϕ_{n-1})`; `block(., ϕ_{n+1})`; `blockfrom(., #_4)`.

The cleanup operation starts by denaturing (warming up) the tube. Immobilized strands are removed from the tube. Next a gel electrophoresis is carried out to separate the longest DNA molecules from the other molecules. Thanks to the conditions we have imposed on inputs to cleanup, the result of this separation is either empty or consists of positive DNA molecules.

7 DNAQL

In this section we define a limited functional programming language, DNAQL, for expressing functions from ℓ-complexes to ℓ-complexes. A crucial feature of DNAQL is that the same program can be applied uniformly to complexes of any particular dimension ℓ. DNAQL is not computationally complete, as it is meant as a query language and not a general-purpose programming language. The language is based on the operations on complexes introduced earlier, and adds to this the following features: some distinguished constants; an emptiness test (if-then-else); let-variable binding; counters that can count up to the dimension of the complex; and a limited for-loop for iterating over a counter.

The syntax of DNAQL is given in Figure 4. Note that expressions can contain two kinds of variables: variables standing for complexes, and counters, ranging from 1 to the dimension. Complex variables can be bound by let-constructs, and counters can be bound by for-constructs. The free (unbound) complex variables of a DNAQL expression stand for its inputs. A DNAQL *program* is a DNAQL expression without free counters. So, in a program, all counters are introduced by for-loops.

$$
\begin{aligned}
\langle expression \rangle &::= \langle complexvar \rangle \mid \langle foreach \rangle \mid \langle if \rangle \mid \langle let \rangle \mid \langle operator \rangle \mid \langle constant \rangle \\
\langle foreach \rangle &::= \textbf{for } \langle complexvar \rangle := \langle expression \rangle \textbf{ iter } \langle counter \rangle \textbf{ do } \langle expression \rangle \\
\langle if \rangle &::= \textbf{if empty}(\langle complexvar \rangle) \textbf{ then } \langle expression \rangle \textbf{ else } \langle expression \rangle \\
\langle let \rangle &::= \textbf{let } x := \langle expression \rangle \textbf{ in } \langle expression \rangle \\
\langle operator \rangle &::= ((\langle expression \rangle) \cup (\langle expression \rangle)) \mid (((\langle expression \rangle) - (\langle expression \rangle))) \\
&\mid \quad \textbf{hybridize}(\langle expression \rangle) \mid \textbf{ligate}(\langle expression \rangle) \\
&\mid \quad \textbf{flush}(\langle expression \rangle) \\
&\mid \quad \textbf{split}(\langle expression \rangle, \langle splitpoint \rangle) \\
&\mid \quad \textbf{block}(\langle expression \rangle, \Sigma) \\
&\mid \quad \textbf{blockfrom}(\langle expression \rangle, \Sigma) \\
&\mid \quad \textbf{blockexcept}(\langle expression \rangle, \langle counter \rangle) \\
&\mid \quad \textbf{cleanup}(\langle expression \rangle) \\
\langle constant \rangle &::= \Sigma^+ \mid (\overline{\Sigma} - \overline{\Lambda}) \, (\overline{\Sigma} - \overline{\Lambda}) \mid \textbf{immob}(\overline{\Sigma}) \\
&\mid \quad \textbf{leftboot} \mid \textbf{rightboot} \mid \textbf{empty} \\
\langle splitpoint \rangle &::= \#_2 \mid \#_3 \mid \#_4 \mid \#_6 \mid \#_8
\end{aligned}
$$

Fig. 4. Syntax of DNAQL

The constants have the following meaning as particular complexes:

- A word $w \in \Sigma^+$ stands for a single, linear, positive strand that spells the word w.

- A two-letter word $\bar{a}\bar{b}$, for $a, b \in \Sigma - \Lambda$, stands for a single, linear, negative strand of length two of the form $1 \xrightarrow{\bar{b}} 2 \xrightarrow{\bar{a}} 3$.
- $\texttt{immob}(\bar{a})$, for $a \in \Sigma$, stands for a single, negative, immobilized edge labeled $\bar{a}$.
- $\texttt{leftboot}$ and $\texttt{rightboot}$ are illustrated in Figure 5.
- $\texttt{empty}$ stands for the empty complex, i.e., the complex with the empty set of nodes.

Fig. 5. Left- and right-boot-shaped complexes

The semantics of a DNAQL expression e is defined relative to a context consisting of a dimension ℓ, an ℓ-*complex assignment* β, and an ℓ-*counter assignment* γ. An ℓ-complex assignment is a mapping from complex variables to ℓ-complexes; an ℓ-counter assignment is a mapping from counters to $\{1, \dots, \ell\}$. Naturally, β must be defined on all free variables of e, and γ must be defined on all free counters of e. Within such a context, the expression can evaluate to an ℓ-complex, denoted by $\llbracket e \rrbracket^{\ell}(\beta, \gamma)$. The semantic rules that define this evaluation are shown in Figure 6. The superscript ℓ has been omitted to reduce clutter. The rules for $\texttt{let}$ and $\texttt{for}$ use the oft-used notation $f[x := u]$ to denote the mapping f updated so that x is mapped to u. Because the operations on complexes are not always defined, the evaluation may fail, so $\llbracket e \rrbracket^{\ell}(\beta, \gamma)$ may be undefined. When e is a program, we denote $\llbracket e \rrbracket(\beta, \emptyset)$ simply by $\llbracket e \rrbracket(\beta)$.

8 Simulation of the Relational Algebra

Let us first recall some basic definitions concerning the relational data model. Basically we assume a universe U of data elements. A *relation schema* R is a finite set of attributes. A *tuple* over R is a mapping from R to U. A *relation* over R is a finite set of tuples over R. A *database schema* is a mapping D on some finite set of relation variables that assigns a relation schema to each relation variable. An *instance* of D is a mapping I on the same set of relation variables that assigns to each relation variable x a relation over $D(x)$.

The syntax of the relational algebra [1] is generated by the following grammar:

$$e ::= x \mid (e \cup e) \mid (e - e) \mid (e \times e) \mid \sigma_{A=B}(e) \mid \widehat{\pi}_A(e) \mid \rho_{A/B}(e) \ .$$

Here, x stands for a relation variable, and A and B stand for attributes. Our version of the relational algebra is slightly nonstandard in that our version of

$$\frac{x \text{ is a complex variable}}{[\![x]\!](\beta,\gamma) = \beta(x)} \qquad \frac{[\![e_1]\!](\beta,\gamma) = C_1 \qquad [\![e_2]\!](\beta,\gamma) = C_2}{[\![e_1 \cup e_2]\!](\beta,\gamma) = C_1 \cup C_2}$$

$$\frac{[\![e_1]\!](\beta,\gamma) = C_1 \qquad [\![e_2]\!](\beta,\gamma) = C_2 \qquad C_1 - C_2 \text{ is well-defined}}{[\![e_1 - e_2]\!](\beta,\gamma) = C_1 - C_2}$$

$$\frac{[\![e']\!](\beta,\gamma) = C'}{[\![\texttt{hybridize}(e')]\!](\beta,\gamma) = \texttt{hybridize}(C')} \qquad \frac{[\![e']\!](\beta,\gamma) = C'}{[\![\texttt{ligate}(e')]\!](\beta,\gamma) = \texttt{ligate}(C')}$$

$$\frac{[\![e']\!](\beta,\gamma) = C'}{[\![\texttt{flush}(e')]\!](\beta,\gamma) = \texttt{flush}(C')} \qquad \frac{[\![e']\!](\beta,\gamma) = C' \qquad \sigma \in \{\#_2, \#_3, \#_4, \#_6, \#_8\}}{[\![\texttt{split}(e',\sigma)]\!](\beta,\gamma) = \texttt{split}(C',\sigma)}$$

$$\frac{[\![e']\!](\beta,\gamma) = C' \qquad \texttt{block}(C',\sigma) \text{ is well-defined}}{[\![\texttt{block}(e',\sigma)]\!](\beta,\gamma) = \texttt{block}(C',\sigma)}$$

$$\frac{[\![e']\!](\beta,\gamma) = C' \qquad \texttt{blockfrom}(C',\sigma) \text{ is well-defined}}{[\![\texttt{blockfrom}(e',\sigma)]\!](\beta,\gamma) = \texttt{blockfrom}(C',\sigma)}$$

$$\frac{[\![e']\!](\beta,\gamma) = C' \qquad i \text{ is a counter} \qquad \texttt{blockexcept}(C',\gamma(i)) \text{ is well-defined}}{[\![\texttt{blockexcept}(e',i)]\!](\beta,\gamma) = \texttt{blockexcept}(C',\gamma(i))}$$

$$\frac{[\![e']\!](\beta,\gamma) = C' \qquad \texttt{cleanup}(C') \text{ is well-defined}}{[\![\texttt{cleanup}(e')]\!](\beta,\gamma) = \texttt{cleanup}(C')}$$

$$\frac{[\![e_1]\!](\beta,\gamma) = C_1 \qquad [\![e_2]\!](\beta[x := C_1],\gamma) = C_2}{[\![\texttt{let } x := e_1 \texttt{ in } e_2]\!](\beta,\gamma) = C_2}$$

$$\frac{[\![e_1]\!](\beta,\gamma) = C_1 \qquad \beta(x) \text{ is the empty complex}}{[\![\texttt{if empty}(x) \texttt{ then } e_1 \texttt{ else } e_2]\!](\beta,\gamma) = C_1}$$

$$\frac{[\![e_2]\!](\beta,\gamma) = C_2 \qquad \beta(x) \text{ is } not \text{ the empty complex}}{[\![\texttt{if empty}(x) \texttt{ then } e_1 \texttt{ else } e_2]\!](\beta,\gamma) = C_2}$$

$$\frac{[\![e_1]\!](\beta,\gamma) = C_0 \qquad [\![e_2]\!](\beta[x := C_{n-1}],\gamma[i := n]) = C_n \text{ for } n = 1,\ldots,\ell}{[\![\texttt{for } x := e_1 \texttt{ iter } i \texttt{ do } e_2]\!](\beta,\gamma) = C_\ell}$$

Fig. 6. Semantics of DNAQL

projection ($\widehat{\pi}$) projects away some given attribute, as opposed to the standard projection which projects on some given subset of the attributes.

The semantics of the relational algebra is well known and we omit a formal definition. A relational algebra expression e can be evaluated in the context of some database instance I that is defined on at least the relation variables occurring in e. When the evaluation succeeds, e evaluates to a relation denoted by $[\![e]\!](I)$.

(The evaluation of a relational algebra operator may fail due to mismatches between the attributes present in the argument relations and the attributes expected by the operator [25].)

We want now to represent relations by complexes. We will store data elements as vectors of atomic value symbols. So formally, we use Λ^* as our universe U. Then a tuple t (relation r, instance I) is said to be of dimension ℓ if all data elements appearing in t (r, I) are strings of length ℓ. Let t be a tuple of dimension ℓ over relation schema R. We may assume a fixed order on the attributes of R, say, $A, \ldots, B$. We then represent t by the following ℓ-complex: (using the constant notation of DNAQL)

$$complex(t) = \#_2 A \#_3 t(A) \#_4 \ldots \#_2 B \#_3 t(B) \#_4 .$$

A relation r of dimension ℓ is then represented by the ℓ-complex $\bigcup \{complex(t) \mid t \in r\}$ which we denote by $complex(r)$. Moreover, a database instance I of dimension ℓ can be represented by the ℓ-complex assignment $complex(I)$ that maps each relation variable x (used as a complex variable) to $complex(I(x))$.

We are now in a position to state our main theorem.

Theorem 1. *Let some database schema D be fixed. Every relational algebra expression e can be translated into a DNAQL program e^{DNA}, such that for each natural number ℓ and for each ℓ-dimensional database instance I over D, if $[\![e]\!](I)$ is defined, then so is $[\![e^{DNA}]\!]^{\ell}(complex(I))$, and*

$$complex \left([\![e]\!] \, (I) \right) = [\![e^{DNA}]\!]^{\ell}(complex(I))$$

(up to isomorphism).

For the proof we introduce a few useful abbreviations. For $a, b \in \Sigma$, we use `blockfromto`(x, a, b) to abbreviate `blockfrom(block`$(x, b), a)$. For attributes A and B, we use $circularize(x, A, B)$ to abbreviate

```
cleanup(ligate(hybridize(
          hybridize(blockfromto(x₆, B, A) ∪ immob(#̄₃))
                                              ∪ #̄₄#̄₂)))  .
```

If x holds a complex of the form $complex(r)$ for some relation r over a schema with first attribute A and last attribute B, then $circularize(x, A, B)$ will equal the complex obtained from x by circularizing every strand [18,4].

The proof now goes by induction on the structure of e.

Union, difference. If e is $e_1 \cup e_2$, then $e^{DNA} = e_1^{DNA} \cup e_2^{DNA}$. If e is $e_1 - e_2$, then $e^{DNA} = e_1^{DNA} - e_2^{DNA}$.

Cartesian product. Let e be of the form $e_1 \times e_2$ with e_1 over relation schema R and e_2 over a disjoint relation schema S. Let A be the first and B be the last attribute of R and let C be the first and D be the last attribute of S. Consider the following DNAQL program e':

> let $x := e_1^{DNA}$ in let $y := e_2^{DNA}$ in
>
>   if empty(x) then empty else if empty(y) then empty else e_4

where e_4 is given by the following:

$$e_4 := \text{cleanup}(\text{split}(\text{split}(\text{blockfromto}(e_5, B, C), \#_2), \#_4))$$

$$e_5 := \textit{circularize}(e_6, A, D)$$

$$e_6 := \text{cleanup}(\text{ligate}(\text{hybridize}[x_6^a \cup x_6^b \cup \overline{\#_5\#_1}]))$$

$$x_6^a := \text{cleanup}(\text{ligate}(\text{hybridize}(x \cup \text{rightboot})))$$

$$x_6^b := \text{cleanup}(\text{ligate}(\text{hybridize}(y \cup \text{leftboot})))$$

Parts e_6^a and e_6^b attach a unique ending (beginning) to the tuples in r (s). The new tuples are added together, in x_6, along with a *sticky bridge* $(\overline{\#_5\#_1})$, resulting in all possible joins of tuples of e_1^{DNA} and e_2^{DNA}. The rest of the expression is concerned with cutting out the $\#_5\#_1$ piece in the middle of the new chains and getting the "old" e_1^{DNA}-tuples back in front of the "new" tuples.

The program e' is not yet quite correct, however, since we assume that the attributes in complex representations of tuples are ordered in lexicographical order. This order may be disrupted by joining tuples from e_1^{DNA} and e_2^{DNA}. Therefore it is necessary to reorder the attribute-value pairs within each tuple resulting from e^{DNA}. Shuffling attribute-value pairs around in a tuple is done using a new technique we call *double bridging*. Instead of using a single sticky bridge, two sticky bridges are hybridized onto one chain. A careful placement of the bridges allows us to cut twice in the chain without separating parts from the chain. Moreover, the two bridges guide the chain into its new conformation.

Next we describe (in outline) a DNAQL program for shuffling some attribute C to the end of a chain. Assume that A is the first attribute, attribute B occurs just in front of C, C is the attribute that we want to move, D occurs exactly after C and E is the last attribute of the chain. The general outline of the program is:

1. Insert the first marker $(\#_6\#_7)$ between attributes B and C.
2. Insert the second marker $(\#_8\#_9)$ between attributes C and D.
3. Insert the third marker $(\#_9\#_1)$ at the end of the chain.
4. Add the two bridges to the mix: $\overline{\#_6\#_8}$ and $\overline{\#_1\#_7}$.
5. Cut at $\#_6$ and $\#_8$ and ligate the resulting complex.
6. Remove the markers from the chains.

An illustration is in Figure 7. A detailed DNAQL program to do these steps will have a similar structure to program e'.

Projection. Let e be of the form $\widehat{\pi}_C(e_1)$, where the relation schema of e_1 is R. Assume that B is the attribute just in front of C and D is the attribute just

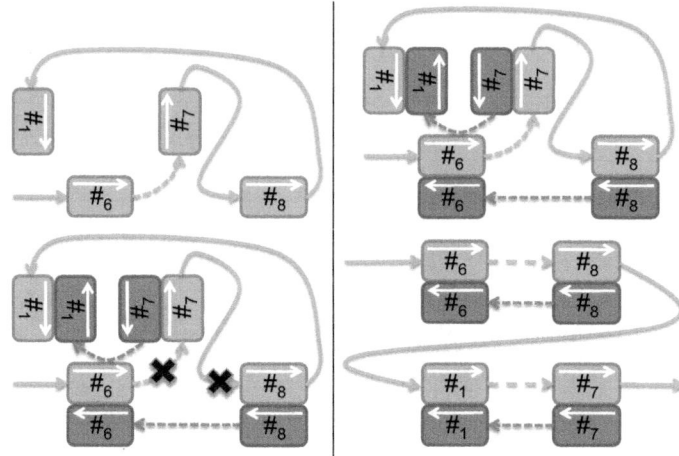

Fig. 7. Illustration of steps 1–3 (top left); step 4 (top right); and step 5 (bottom left, which simplifies to bottom right) described in the proof of simulation of Cartesian product

after attribute C. In the case that attribute C is the first attribute of the relation schema R, B is the last attribute of R. Likewise in the case that attribute C is the last attribute of R, then D is the first attribute of R. We thus perceive R to be circular. Assume that A and E are the first resp. last attribute of R.

We define e^{DNA} as the following program:

$$\texttt{let } x := e_1^{DNA} \texttt{ in if empty}(x) \texttt{ then empty else } f_1$$

where

$$f_1 := \texttt{cleanup(split(blockfromto(cleanup(ligate}(f_2)), E, A), \#_4))$$
$$f_2 := circularize(f_3, D, B)$$
$$f_3 := \texttt{cleanup(split(blockfromto(cleanup(ligate}(f_4)), B, D), \#_4))$$
$$f_4 := circularize(x, A, E)$$

Renaming. Let e be of the form $\rho_{C/F}(e_1)$, where R is the relation schema of e_1. Simulating renaming involves the following steps:

1. Rotate the chains to get attribute C at the start of each chain.
2. Cut the attribute from the chain, leaving the values of C on the chain.
3. Add the F attribute using *stickers*.
4. Rotate the chains again to get the first attribute at the start of each chain.

Assume that attribute B occurs just in front of C, D just after C, A is the first attribute of R and E is the last attribute. Then e^{DNA} is the following program:

$$\texttt{let } x := e_1^{DNA} \texttt{ in if empty}(x) \texttt{ then empty else } f_1$$

where

$f_1 := \texttt{cleanup}(\texttt{split}(\texttt{blockfromto}(f_2, E, A), \#_4))$

$f_2 := \texttt{cleanup}(\texttt{ligate}(\texttt{hybridize}[f_3 \cup \#_2 F \cup \overline{\#_4 \#_2} \cup \overline{F \#_3}]))$

$f_3 := \texttt{hybridize}(\texttt{split}(\texttt{blockfromto}(f_4, B, D), \#_3) \cup \texttt{immob}(\overline{\#_3}))$

$f_4 := \texttt{cleanup}(\texttt{split}(\texttt{blockfromto}(\texttt{cleanup}(\texttt{ligate}(f_5)), B, D), \#_2))$

$x_5 := circularize(x, A, E)$

This program is not yet fully correct as attribute F may need to be shuffled into the right place. This can be done by repeatedly applying the shuffle procedure described in the case of cartesian product.

Selection. Let e be of the form $\sigma_{B=D}(e_1)$, where R is the relation schema of e_1. Translating the selection operator requires the most complicated expressions thus far. Assume that relation schema R has A as its first attribute, C following directly behind B, E following directly after D and F the last attribute of the schema. The Λ is fixed. The number of atomic value symbols is thus a constant; we denote them by v_1 to v_n. Note $A = B$, or $C = D$ or $D = E = F$ is possible; the program will still function correctly.

We define e^{DNA} as follows:

$\texttt{let } x := e_1^{DNA} \texttt{ in if empty}(x) \texttt{ then empty else for } x_s := x \texttt{ iter i do } e'$

where

$e' := \texttt{cleanup}(\texttt{split}(\texttt{blockfromto}($
$\qquad \texttt{let } x_c := circularize(x_s, A, F) \texttt{ in } e'', F, A), \#_4))$

$e'' := select_{v_1}^D(select_{v_1}^B(x_c)) \cup \cdots \cup select_{v_n}^D(select_{v_n}^B(x_c))$

$select_a^B(x') := \texttt{cleanup}(\texttt{flush}(\texttt{hybridize}(e_1^a(x'))))$

$e_1^a(x') := \texttt{blockexcept}(\texttt{blockfromto}(x', B, C), \texttt{i}) \cup immob(\overline{a})$

$select_a^F(x') := \texttt{cleanup}(\texttt{flush}(\texttt{hybridize}(e_2^a(x'))))$

$e_2^a(x') := \texttt{blockexcept}(\texttt{blockfromto}(x', D, E), \texttt{i}) \cup immob(\overline{a})$

9   Concluding Remarks

Many interesting questions remain open. A first issue is that an arbitrary DNAQL program may not evaluate on all possible inputs. We would like to have a type system by which programs can be statically typechecked to be safe on inputs of given types.

We would also like to better understand the expressive power of DNAQL. The relational algebra provides a lower bound on this expressive power. What is an upper bound? Can the semantics of DNAQL be defined in first-order logic? What is the computational complexity of DNAQL? Also, are all operations and constructs of DNAQL really primitive in the language, or can some of them be simulated using the others?

Another interesting issue is the relationship between DNAQL and graph grammars. Furthermore, we could consider extensions, or restrictions, of DNAQL, just this has been done for the relational algebra. Extensions can lead to greater expressive power, while restrictions may lead to decidable static verification problems, such as testing the equivalence of DNAQL programs.

Finally, while we have gone to great efforts to design an abstraction that is as plausible as possible, of course, it would be great if it could be experimentally verified if DNAQL is workeable for practical DNA computing.

References

1. Abiteboul, S., Hull, R., Vianu, V.: Foundations of Databases. Addison-Wesley (1995)
2. Adleman, L.M.: Molecular computation of solutions to combinatorial problems. Science 226, 1021–1024 (1994)
3. Amos, M.: Theoretical and Experimental DNA Computation. Springer, Heidelberg (2005)
4. Arita, M., Hagiya, M., Suyama, A.: Joining and rotating data with molecules. In: Proceedings 1997 IEEE International Conference on Evolutionary Computation, pp. 243–248 (1997)
5. Boneh, D., Dunworth, C., Lipton, R.J., Sgall, J.: On the computational power of DNA. Discrete Applied Mathematics 71, 79–94 (1996)
6. Cardelli, L.: Strand algebras for DNA computing. In: Deaton and Suyama [9], pp. 12–24
7. Chen, J., Deaton, R.J., Wang, Y.-Z.: A DNA-based memory with in vitro learning and associative recall. Natural Computing 4(2), 83–101 (2005)
8. Condon, A.E., Corn, R.M., Marathe, A.: On combinatorial DNA word design. Journal of Computational Biology 8(3), 201–220 (2001)
9. Deaton, R., Suyama, A. (eds.): DNA 15. LNCS, vol. 5877. Springer, Heidelberg (2009)
10. Diatchenko, L., Lau, Y.F., et al.: Suppression subtractive hybridization: a method for generating differentially regulated or tissue-specific cDNA probes and libraries. Proceedings of the National Academy of Sciences 93(12), 6025–6030 (1996)
11. Dirks, R.M., Pierce, N.A.: Triggered amplification by hybridization chain reaction. Proceedings of the National Academy of Sciences 101(43), 15275–15278 (2004)
12. Immerman, N.: Descriptive Complexity. Springer, Heidelberg (1999)
13. Liu, Q., Wang, L., et al.: DNA computing on surfaces. Nature 403, 175–179 (1999)
14. Lyngsø, R.B.: Complexity of Pseudoknot Prediction in Simple Models. In: Díaz, J., Karhumäki, J., Lepistö, A., Sannella, D. (eds.) ICALP 2004. LNCS, vol. 3142, pp. 919–931. Springer, Heidelberg (2004)
15. Majumder, U., Reif, J.H.: Design of a biomolecular device that executes process algebra. In: Deaton and Suyama [9], pp. 97–105
16. Paun, G., Rozenberg, G., Salomaa, A.: DNA Computing. Springer, Heidelberg (1998)
17. Ramakrishnan, R., Gehrke, J.: Database Management Systems. McGraw-Hill (2002)

18. Reif, J.H.: Parallel biomolecular computation: models and simulations. Algorithmica 25(2-3), 142–175 (1999)
19. Reif, J.H., LaBean, T.H., Pirrung, M., Rana, V.S., Guo, B., Kingsford, C., Wickham, G.S.: Experimental Construction of Very Large Scale DNA Databases with Associative Search Capability. In: Jonoska, N., Seeman, N.C. (eds.) DNA 2001. LNCS, vol. 2340, pp. 231–247. Springer, Heidelberg (2002)
20. Rozenberg, G., Spaink, H.: DNA computing by blocking. Theoretical Computer Science 292, 653–665 (2003)
21. Sager, J., Stefanovic, D.: Designing Nucleotide Sequences for Computation: A Survey of Constraints. In: Carbone, A., Pierce, N.A. (eds.) DNA 2005. LNCS, vol. 3892, pp. 275–289. Springer, Heidelberg (2006)
22. Sakamoto, K., et al.: State transitions by molecules. Biosystems 52, 81–91 (1999)
23. Seelig, G., Soloveichik, D., Zhang, D.Y., Winfree, E.: Enzyme-free nucleic acid logic circuits. Science 315(5805), 1585–1588 (2006)
24. Shortreed, M.R., et al.: A thermodynamic approach to designing structure-free combinatorial DNA word sets. Nucleic Acids Research 33(15), 4965–4977 (2005)
25. Van den Bussche, J., Van Gucht, D., Vansummeren, S.: A crash course in database queries. In: Proceedings 26th ACM Symposium on Principles of Database Systems, pp. 143–154. ACM Press (2007)
26. Yamamoto, M., Kita, Y., Kashiwamura, S., Kameda, A., Ohuchi, A.: Development of DNA Relational Database and Data Manipulation Experiments. In: Mao, C., Yokomori, T. (eds.) DNA12. LNCS, vol. 4287, pp. 418–427. Springer, Heidelberg (2006)

Efficient and Accurate Haplotype Inference by Combining Parsimony and Pedigree Information

Ana Graça[1], Inês Lynce[1], João Marques-Silva[2], and Arlindo L. Oliveira[1]

[1] INESC-ID/IST, Technical University of Lisbon
{assg,ines}@sat.inesc-id.pt, aml@inesc-id.pt
[2] CSI/CASL, University College Dublin
jpms@ucd.ie

Abstract. Existing genotyping technologies have enabled researchers to genotype hundreds of thousands of SNPs efficiently and inexpensively. Methods for the imputation of non-genotyped SNPs and the inference of haplotype information from genotypes, however, remain important, since they have the potential to increase the power of statistical association tests. In many cases, studies are conducted in sets of individuals where the pedigree information is relevant, and can be used to increase the power of tests and to decrease the impact of population structure on the obtained results. This paper proposes a new Boolean optimization model for haplotype inference combining two combinatorial approaches: the Minimum Recombinant Haplotyping Configuration (MRHC), which minimizes the number of recombinant events within a pedigree, and the Haplotype Inference by Pure Parsimony (HIPP), that aims at finding a solution with a minimum number of distinct haplotypes within a population. The paper also describes the use of well-known techniques, which yield significant performance gains. Concrete examples include symmetry breaking, identification of lower bounds, and the use of an appropriate constraint solver. Experimental results show that the new PedRPoly model is competitive both in terms of accuracy and efficiency.

Keywords: Haplotype inference, Boolean optimization.

1 Introduction

The majority of complex diseases are influenced by both environmental and genetic factors. Existing technologies have enabled researchers to genotype hundreds of thousands of single nucleotide polymorphisms (SNPs) in a single run. Data obtained with genotyping or sequencing technologies for thousands of individuals will be available in the near future. This will enable researchers to conduct whole genome association studies in an unprecedented scale to detect increasingly subtle and more complex associations between genomes and diseases [33].

K. Horimoto, M. Nakatsu, and N. Popov (Eds.): ANB 2011, LNCS 6479, pp. 38–56, 2012.
© Springer-Verlag Berlin Heidelberg 2012

Despite these technological advances, existing technologies still generate genotypes obtained from the conflation of two haplotypes on homologous chromosomes. Hence, haplotypes must be inferred computationally using the experimentally identified genotypes. Inference of the haplotypes is made possible by the fact that, in many cases, there exists a strong correlation between the allele present in a particular SNP and other nearby sites. A given combination of alleles in one chromosome is termed a haplotype, and the deviation from independence that exists between alleles is known as linkage disequilibrium.

Haplotype inference from genotype data remains an important and challenging task. The identification of haplotypes allows to develop haplotype-based association studies [5]. In addition, most imputation methods require the haplotype data [30].

Many haplotype inference methods apply to unrelated individuals. Nonetheless, pedigree information is available in many studies and can be used to improve the results of inference methods. Combinatorial methods for haplotype inference have been shown to be practical and relevant, either for phasing families [20] or unrelated individuals [10]. Recently, a study comparing the haplotype inference methods using pedigrees and unrelated individuals [21] concluded that taking into consideration both pedigree and population information leads to improvements on the precision of haplotype inference methods.

This paper proposes the combination of two well-known haplotype inference methods: pure parsimony, which aims at finding a solution that uses the minimum possible number of distinct haplotypes, and the minimum recombinant approach used to phase individuals organized in pedigrees by minimizing the number of recombination events within each pedigree. The resulting haplotype inference model, PedRPoly, is shown to be quite competitive and more accurate than the existing methods for haplotype inference from pedigrees, in particular, using the minimum recombinant approach [20]. Note that a simpler and fairly inefficient version of this model was first outlined in [9].

The models developed in this paper represent challenging combinatorial optimization problems, which can be viewed as a special case of multi-objective optimization. The solution methods for these problems combine techniques often used in the fields of operations research and artificial intelligence. Furthermore, the proposed models yield accuracy results that conclusively outperform the minimum recombinant approach. This paper describes the first PedRPoly algorithm which can actually be used in practice, since it is both efficient and accurate.

This paper is organized in two main parts. The first part describes the PedRPoly model, which combines the pure parsimony and the minimum recombinant approaches. This model is enhanced with several constraint modeling techniques. These techniques aim at improving the efficiency of the method and include the identification of lower bounds, symmetry breaking and a heuristic sorting technique. The second part conducts a comprehensive experimental evaluation of the accuracy and efficiency of PedRPoly on a large set of instances. The experimental evaluation was also used to select the best performing constraint solver, among a representative number of solvers.

2   Haplotype Inference

Variations in the DNA sequence are the basis for evolution. *Single Nucleotide Polymorphisms* (SNPs) are the most common variations between human beings, and occur when a nucleotide base (A, T, C, G) is changed to other nucleotide base at a single DNA position. Moreover, the mutant type nucleotide must be represented in a significant percentage of the population (normally 1%). An example of a SNP is the mutation of the DNA sequence AC**TT**GAC to AC**AT**GAC, where the third nucleotide is changed from T to A. DNA is organized in structures called chromosomes. Chromosomes contain the genetic information coded in different type of substructures, of which the best known are the genes. A gene is a sequence of DNA bases which encodes a specific protein.

A given combination of SNPs in a single chromosome is called a *haplotype*. Moreover, SNPs within a haplotype tend to be inherited together. The deviation from independence that exists between SNPs is known as *linkage disequilibrium* (LD).

Diploid organisms, such as human beings, have pairs of homologous chromosomes, with each chromosome in a pair inherited from a single parent. In practice, experimental technology is only able to obtain genotypes, which correspond to the conflated data of two haplotypes on homologous chromosomes. The haplotype inference problem consists in obtaining the set of haplotype pairs which originated a given set of genotypes.

Considering that the assumptions underlying the infinite-site model [14] are valid, we may assume that each SNP can only have two values (called alleles). Each haplotype can therefore be represented by a binary string, with size $m \in \mathbb{N}$, where 0 represents the wild type allele and 1 represents the mutant type allele. Each site of the haplotype h_i is represented by h_{ij} ($1 \leq j \leq m$). In addition, each genotype is represented by a string, with size m, over the alphabet $\{0, 1, 2\}$, and each site of the genotype g_i is represented by g_{ij}. Each genotype is explained by two haplotypes. A genotype g_i is explained by a pair of haplotypes (h_i^a, h_i^b), which is represented by $g_i = h_i^a \oplus h_i^b$, if

$$g_{ij} = \begin{cases} h_{ij}^a & \text{if } h_{ij}^a = h_{ij}^b \\ 2 & \text{if } h_{ij}^a \neq h_{ij}^b \end{cases}.$$

A genotype site g_{ij} with either value 0 or 1 is a homozygous site (the same allele is inherited from both parents), whereas a site with value 2 is a heterozygous site (different alleles are inherited from each parent).

Definition 1. *(Haplotype Inference) Given a set $\mathcal{G}$ of n genotypes, each with size m, the haplotype inference problem consists in finding a set of haplotypes $\mathcal{H}$, such that each genotype $g_i \in \mathcal{G}$ is explained by two haplotypes $h_i^a, h_i^b \in \mathcal{H}$.*

Observe that for each genotype g with k heterozygous sites, there are 2^{k-1} nonordered pairs of haplotypes that can explain g. For example, genotype $g = 022$ can be explained either by haplotypes $(000, 011)$ or by haplotypes $(001, 010)$.

Most often genotyping procedures leave a percentage of missing data, i.e. genotype positions with unknown values. To represent missing sites, the alphabet of the genotypes is extended to $\{0, 1, 2, ?\}$.

Pedigrees. Pedigree data adds new relevant information to the haplotype inference problem. A pedigree refers to the genealogical tree which allows studying the inheritance of genes within a family. The *Mendelian laws of inheritance* are well established assumptions and, in particular, state that all sites in a single haplotype are inherited from a single parent, assuming there are no mutations within a pedigree. We assume that haplotype h^a is inherited from the father and h^b is inherited from the mother. Nonetheless, a recombination may happen. A recombination occurs when two haplotypes of a parent are mixed together and the recombinant is inherited by the child. For example, suppose a father has the haplotype pair $(000, 111)$ and the haplotype that he passed on to his child is 100. Here, one recombination event must have occurred: haplotypes 000 and 111 got shuffled and originated a new haplotype $h = 100$. Therefore, the child has inherited the first allele from the paternal grandmother, while second and third alleles were inherited from the paternal grandfather. In a pedigree, an individual is a *founder* if he does not have parents on the pedigree (and a *non-founder* if he has both parents on the pedigree).

2.1   Minimum Recombinant Haplotype Configuration

Most rule-based haplotype inference methods for pedigrees assume no recombination within each pedigree [35, 37, 22]. The assumption of no recombination is valid in many cases because recombination events are rare in DNA regions with high linkage disequilibrium. Nonetheless, this assumption can be violated even for some dense markers [19]. Therefore, a more realistic approach consists in minimizing the number of recombinations within pedigrees [13, 31, 20].

Definition 2. *The minimum recombinant haplotype configuration (MRHC) problem aims at finding a haplotype inference solution for a pedigree which minimizes the number of required recombination events.*

For example, suppose the father has the genotype $g_1 = 202$, the mother has the genotype $g_2 = 212$ and the child has the genotype $g_3 = 222$. One possible solution to the haplotype inference problem is $g_1 = 001 \oplus 100$, $g_2 = 010 \oplus 111$ and $g_3 = 101 \oplus 010$. However, this solution implies that one recombination has occurred, because the child has not inherited an integral haplotype from his father, but a mixture of his paternal grandparents haplotypes. A different solution to this example admits no recombination and, therefore, is a MRHC solution: $g_1 = 000 \oplus 101$, $g_2 = 010 \oplus 111$ and $g_3 = 000 \oplus 111$.

In general, there can be a significant number of MRHC solutions to the same problem. For instance, $g_1 = 000 \oplus 101$, $g_2 = 010 \oplus 111$, $g_3 = 101 \oplus 010$ is another 0-recombinant solution for the previous example, i.e. another MRHC solution.

The MRHC problem has been shown to be a NP-hard [19, 25] problem. The PedPhase tool [20] implements an integer linear programming (ILP) model for MRHC with missing alleles.

2.2 Haplotype Inference by Pure Parsimony

The haplotype inference by pure parsimony problem consists in finding a solution to the haplotype inference problem which minimizes the number of distinct haplotypes [12].

Natural phenomena tend to be explained parsimoniously, using the minimum number of required entities. The haplotype inference by pure parsimony approach is also biologically motivated by the fact that individuals from the same population have the same ancestors and mutations do not occur often. Moreover, it is also well-known that the number of haplotypes in a population is much smaller than the number of genotypes [34].

Definition 3. *The haplotype inference by pure parsimony (HIPP) approach aims at finding a minimum-cardinality set of haplotypes $\mathcal{H}$ that can explain a given set of genotypes $\mathcal{G}$.*

For example, consider the set of genotypes $\mathcal{G} = \{g_1,\ g_2,\ g_3\} = \{202,\ 212,\ 222\}$. There are solutions using 6 different haplotypes: $\mathcal{H}_1 = \{101,\ 000,\ 111,\ 010,\ 001,\ 110\}$, such that $g_1 = 101 \oplus 000$, $g_2 = 111 \oplus 010$ and $g_3 = 001 \oplus 110$. However the HIPP solution only requires 4 distinct haplotypes: $\mathcal{H}_2 = \{101,\ 000,\ 111,\ 010\}$ such that $g_1 = 101 \oplus 000$, $g_2 = 111 \oplus 010$ and $g_3 = 000 \oplus 111$.

The HIPP problem is NP-hard [16]. RPoly [10] is a state-of-the-art solver implementing a 0-1 ILP model for solving the HIPP problem.

3 The PedRPoly Model

This section describes the *minimum recombinant maximum parsimony* model, denoted as PedRPoly model. In practice, the PedRPoly model is a combination of the MRHC PedPhase model [20] and the HIPP RPoly model [10].

Definition 4. *Given sets of pedigrees from the same population, the minimum recombinant maximum parsimony (MRMP) model aims at finding a haplotype inference solution which first minimizes the number of recombination events within pedigrees and then minimizes the number of distinct haplotypes used.*

Figure 1 illustrates two trios (mother $\bigcirc$, father $\square$ and child $\Diamond$) from two families A and B with the corresponding genotypes. The figure includes three haplotype inference solutions. Solution 1 is a 0-recombinant solution with 7 distinct haplotypes $\{100, 101, 000, 111, 011, 001, 110\}$. Solution 2 is a 1-recombinant solution (there is one recombination event in family B) using 5 distinct haplotypes $\{100, 101, 000, 111, 011\}$. Solution 3 is a 0-recombinant solution using 5 distinct haplotypes $\{100, 101, 000, 111, 011\}$.

According to the PedRPoly model, solution 3 is preferred to the other solutions. Solution 3 is both a MRHC and a HIPP solution. Consequently, solution 3 is a MRMP solution. If there exists no solution that minimizes both criteria, then preference is given to the MRHC criterion. Hence, the MRHC solution which uses the smallest number of distinct haplotypes would be chosen.

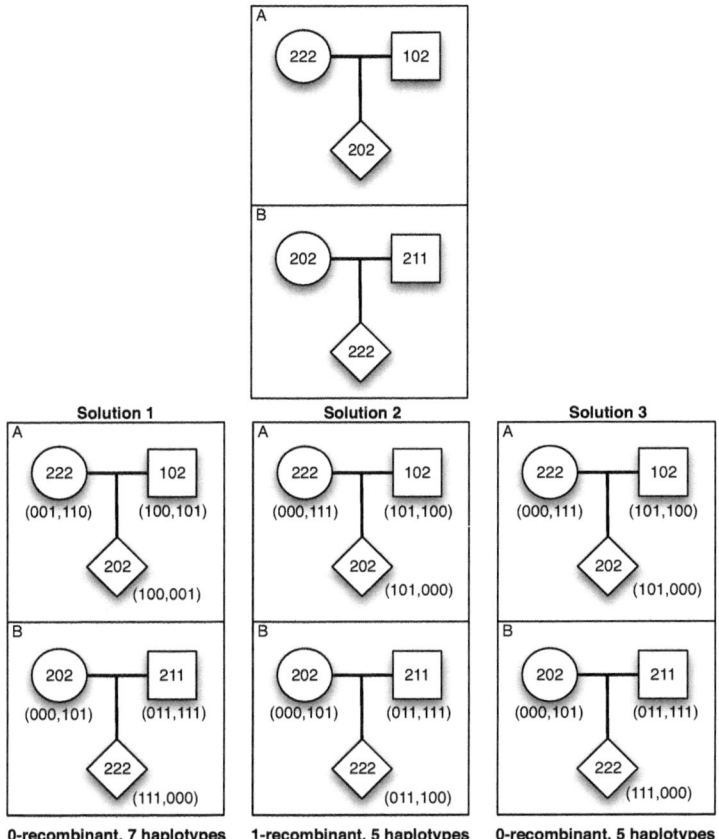

Fig. 1. Solutions for haplotype inference with two trios

The minimum recombinant maximum parsimony model combines the Ped-Phase and the RPoly models, in a new 0-1 integer linear programming model, so called PedRPoly. A 0-1 integer linear programming problem aims at finding a Boolean assignment to the variables which optimizes the value of a given cost function, subject to a set of linear constraints. The cost function and the general constraints of the PedRPoly model are detailed in Table 1, which is described in the following paragraphs.

Following the RPoly model, PedRPoly associates two haplotypes, h_i^a and h_i^b, with each genotype g_i, and these haplotypes are required to explain g_i. Moreover, PedRPoly associates a variable t_{ij} with each heterozygous site g_{ij}, such that $t_{ij} = 1$ indicates that the mutant value was inherited from the father ($h_{ij}^a = 1$) and the wild value was inherited from the mother ($h_{ij}^b = 0$) whereas $t_{ij} = 0$ indicates that the wild value was inherited from the father ($h_{ij}^a = 0$) and the mutant value was inherited from the mother ($h_{ij}^b = 1$). In addition, PedRPoly associates two variables with each missing site. Variable t_{ij}^a is associated with the

Table 1. The PedRPoly Model

minimize:	$((2n+1) \times \sum_{\text{non-founder } i} \sum_{j=1}^{m-1}(r_{ij}^1 + r_{ij}^2)) + \sum_{i=1}^{n}(u_i^a + u_i^b)$	
subject to:		
Equation	Constraint Mendelian laws of inheritance rules (Table 2)	Indexes
		$l \in \{1,2\}$
(1)	$-r_{ij}^l + g_{ij}^l - g_{ij+1}^l \leq 0$ $-r_{ij}^l - g_{ij}^l + g_{ij+1}^l \leq 0$	$1 \leq i \leq n,\ i$ non-founder $1 \leq j \leq m-1$ $p,q \in \{a,b\}$
(2)	$\neg(R \Leftrightarrow S) \Rightarrow x_{ik}^{pq}$ (Table 3)	$1 \leq k < i \leq n$ $1 < i \leq n$
(3)	$\sum_{k<i\,;\,q\in\{a,b\}} x_{ik}^{pq} - u_i^p \leq 2i - 3$	$p \in \{a,b\}$

paternal haplotype site h_{ij}^a, whereas variable t_{ij}^b is associated with the maternal haplotype site h_{ij}^b. The values of h_i^a and h_i^b at homozygous sites are implicitly assumed.

The grandparental origin of each site of the haplotypes must be considered when analyzing recombination events within pedigrees. Following the MRHC PedPhase model, for each non-founder individual i and site j, two variables are defined: g_{ij}^1 and g_{ij}^2. The assignment $g_{ij}^1 = 0$ means that the paternal allele of individual i at site j (i.e. h_{ij}^a) comes from the paternal grandfather, and $g_{ij}^1 = 1$ means that h_{ij}^a comes from the paternal grandmother, i.e.

$$g_{ij}^1 = \begin{cases} 0 \text{ if } h_{ij}^a = h_{f(i)j}^a \\ 1 \text{ if } h_{ij}^a = h_{f(i)j}^b \end{cases},$$

where $f(i)$ corresponds to the father of individual i. In a similar way, $g_{ij}^2 = 0$ ($g_{ij}^2 = 1$) means that the maternal allele of individual i at site j comes from the maternal grandfather (grandmother), i.e.

$$g_{ij}^2 = \begin{cases} 0 \text{ if } h_{ij}^b = h_{m(i)j}^a \\ 1 \text{ if } h_{ij}^b = h_{m(i)j}^b \end{cases},$$

where $m(i)$ corresponds to the mother of individual i.

Constraints to ensure that the Mendelian laws of inheritance are satisfied are defined in Table 2. Note that PedRPoly only associates variables with heterozygous and missing sites (inspired by RPoly), while PedPhase also associates variables with homozygous sites. The new definition of variables associated with sites requires the redefinition of the constraints related with Mendelian laws. For instance, consider the first constraint of Table 2, $t_{f(i)j} \Leftrightarrow g_{ij}^1$, for the case

Table 2. Mendelian laws of inheritance rules regarding variables g_{ij}^1. (The constraints involving variables g_{ij}^2 are defined similarly. $f(i)$ corresponds to the father of i. $1 \leq i \leq n$, i non-founder, $1 \leq j \leq m$.)

Condition	Constraint
$g_{ij} = 0 \wedge g_{f(i)j} = 2$	$t_{f(i)j} \Leftrightarrow g_{ij}^1$
$g_{ij} = 0 \wedge g_{f(i)j} = ?$	$(g_{ij}^1 \vee \neg t_{f(i)j}^a) \wedge (\neg g_{ij}^1 \vee \neg t_{f(i)j}^b)$
$g_{ij} = 1 \wedge g_{f(i)j} = 2$	$t_{f(i)j} \Leftrightarrow \neg g_{ij}^1$
$g_{ij} = 1 \wedge g_{f(i)j} = ?$	$(g_{ij}^1 \vee t_{f(i)j}^a) \wedge (\neg g_{ij}^1 \vee t_{f(i)j}^b)$
$g_{ij} = 2 \wedge g_{f(i)j} = 0$	$\neg t_{ij}$
$g_{ij} = 2 \wedge g_{f(i)j} = 1$	t_{ij}
$g_{ij} = 2 \wedge g_{f(i)j} = 2$	$(g_{ij}^1 \vee t_{ij} \vee \neg t_{f(i)j}) \wedge (g_{ij}^1 \vee \neg t_{ij} \vee t_{f(i)j}) \wedge$ $(\neg g_{ij}^1 \vee t_{ij} \vee t_{f(i)j}) \wedge (\neg g_{ij}^1 \vee \neg t_{ij} \vee \neg t_{f(i)j})$
$g_{ij} = 2 \wedge g_{f(i)j} = ?$	$(g_{ij}^1 \vee t_{ij} \vee \neg t_{f(i)j}^a) \wedge (g_{ij}^1 \vee \neg t_{ij} \vee t_{f(i)j}^a) \wedge$ $(\neg g_{ij}^1 \vee t_{ij} \vee \neg t_{f(i)j}^b) \wedge (\neg g_{ij}^1 \vee \neg t_{ij} \vee t_{f(i)j}^b)$
$g_{ij} = ? \wedge g_{f(i)j} = 0$	$\neg t_{ij}^a$
$g_{ij} = ? \wedge g_{f(i)j} = 1$	t_{ij}^a
$g_{ij} = ? \wedge g_{f(i)j} = 2$	$(g_{ij}^1 \vee t_{ij}^a \vee \neg t_{f(i)j}) \wedge (g_{ij}^1 \vee \neg t_{ij}^a \vee t_{f(i)j}) \wedge$ $(\neg g_{ij}^1 \vee t_{ij}^a \vee t_{f(i)j}) \wedge (\neg g_{ij}^1 \vee \neg t_{ij}^a \vee \neg t_{f(i)j})$
$g_{ij} = ? \wedge g_{f(i)j} = ?$	$(g_{ij}^1 \vee t_{ij}^a \vee \neg t_{f(i)j}^a) \wedge (g_{ij}^1 \vee \neg t_{ij}^a \vee t_{f(i)j}^a) \wedge$ $(\neg g_{ij}^1 \vee t_{ij}^a \vee \neg t_{f(i)j}^b) \wedge (\neg g_{ij}^1 \vee \neg t_{ij}^a \vee t_{f(i)j}^b)$

$g_{ij} = 0$ and $g_{f(i)j} = 2$. Clearly, if $t_{f(i)j} = 1$ (representing that individual $f(i)$ has inherited value 1 from his father and value 0 from his mother) then $g_{ij}^1 = 1$ (representing that individual i must have inherited the value 0 from his paternal grandmother) and conversely.

In addition, in order to allow counting the number of recombinations, the model defines new variables r. For each non-founder individual i, variable r_{ij}^1 (r_{ij}^2) is assigned value 1 if a recombination took place at site j, to create the paternal (maternal) haplotype of individual i. Thus, $r_{ij}^l = 1$ if $g_{ij}^l \neq g_{ij+1}^l$, for $l \in \{1, 2\}$ and $1 \leq j \leq m - 1$, which is ensured by constraints (1) in Table 1. Here, another simplification to the original MRHC is considered. Actually, in the PedPhase model, $r_{ij}^l = 1$ *if and only if* $g_{ij}^l \neq g_{ij+1}^l$. Observe that an implication, instead of an equivalence, is sufficient for correctness and reduces in half the number of these constraints.

Moreover, the model defines variables to count the number of distinct haplotypes used. Let x_{ik}^{pq}, with $p, q \in \{a, b\}$ and $1 \leq k < i \leq n$, be 1 if haplotype p of genotype g_i (h_i^p) and haplotype q of genotype g_k (h_k^q) are different. The conditions on the x_{ik}^{pq} variables are based on the values of variables t_{ij} and t_{kj} for heterozygous sites and of variables t_{ij}^a, t_{ij}^b, t_{kj}^a and t_{kj}^b for missing sites, and are described by equations (2) in Table 1.

Furthermore, the model needs variables u to denote when one of the haplotypes, associated with a given genotype, is different from all previous haplotypes. Hence, u_i^p, with $p \in \{a, b\}$ and $1 \leq i \leq n$, is 1 if haplotype p of genotype g_i

Table 3. Definition of predicates R and S, accordingly to index values

Condition	Constraint
$g_{ij} \neq 2 \wedge g_{kj} = 2$	$R = (g_{ij} \Leftrightarrow (q \Leftrightarrow a))$ and $S = t_{kj}$
$g_{kj} \neq 2 \wedge g_{ij} = 2$	$R = (g_{kj} \Leftrightarrow (p \Leftrightarrow a))$ and $S = t_{ij}$
$g_{ij} = 2 \wedge g_{kj} = 2$	$R = (p \Leftrightarrow q)$ and $S = (t_{ij} \Leftrightarrow t_{kj})$
$g_{ij} =? \wedge g_{kj} \notin \{2,?\}$	$R = t_{ij}^{p}$ and $S = g_{kj}$
$g_{kj} =? \wedge g_{ij} \notin \{2,?\}$	$R = t_{kj}^{q}$ and $S = g_{ij}$
$g_{ij} =? \wedge g_{kj} = 2$	$R = (q \Leftrightarrow a)$ and $S = (t_{ij}^{p} \Leftrightarrow t_{kj})$
$g_{kj} =? \wedge g_{ij} = 2$	$R = (p \Leftrightarrow a)$ and $S = (t_{kj}^{q} \Leftrightarrow t_{ij})$
$g_{ij} =? \wedge g_{kj} =?$	$R = t_{ij}^{p}$ and $S = t_{kj}^{q}$

is different from all previous haplotypes. Then, the conditions on the u_i^p variables are based on the conditions for the x_{ik}^{pq} variables, with $1 \leq k < i$ and $q \in \{a, b\}$. These conditions are described by equations (3) in Table 1.

Finally, the cost function consists in minimizing the number of recombination events and the number of distinct haplotypes, which are, respectively, given by the sum of variables r and u,

$$minimize \quad ((2n + 1) \times \sum_{\substack{(\text{non-founder } i)}} \sum_{j=1}^{m-1} (r_{ij}^1 + r_{ij}^2)) + \sum_{i=1}^{n} (u_i^a + u_i^b). \quad (1)$$

Given that higher importance is given to the minimum recombinant criterion, a larger weight is given to the number of recombinations. Note that $2n$ is a trivial upper bound on the number of haplotypes in the solution, and therefore giving weight $2n + 1$ to the number of recombinations implies that a MRHC solution is always preferred. The idea of giving more weight to the number of recombinations is biological motivated by the fact that recombination events within haplotypes in a pedigree are rare. Moreover, note that a larger number of recombinants suggests a larger number of haplotypes. In general, a recombination event generates a new haplotype, whereas without recombination, the haplotypes of the child are exact copies of the parents' haplotypes. Nonetheless, different weights w, $1 < w < 2n + 1$, were also tried but did not lead to improvements neither on accuracy or efficiency.

Finally, we would like to point out that a two-step approach which obtains all MRHC solutions first and then picks the solution with the smallest number of haplotypes would not be practical. The number of MRHC solutions is, in general, significantly large, specially with higher missing rates, and usually it is not feasible to compute all solutions. In addition, note that the number of all MRHC solutions is the product of the number of MRHC solutions for each pedigree. On the other hand, the minimum recombinant maximum parsimony criterion reduces the search space and results confirm that it produces more accurate results.

4   Improving Efficiency

This section describes three improvements on the original PedRPoly model, which contribute for an efficient haplotype inference model. The practical contribution of each technique is detailed in Section 5.1.

4.1   Lower Bounds

The integration of lower bounds is a modeling technique implemented previously in other approaches [26, 11]. The algorithms for computing lower bounds rely on information regarding (in)compatible genotypes. Two genotypes are declared *compatible* if does *not* exist a site for which one genotype has value 0 and the other genotype has value 1. Otherwise, the genotypes are *incompatible*. Clearly, two incompatible genotypes cannot be explained by the same haplotypes. Given the incompatibility relation we can create an *incompatibility* graph I, where each vertex is a genotype, and two vertexes are linked with an edge if they are incompatible. Suppose I has a clique of size k. Hence, the number of required haplotypes is at least $2 \cdot k - \sigma$, where σ is the number of genotypes in the clique which do not have heterozygous sites.

In addition, an analysis of the structure of the genotypes allows the lower bound to be further increased. The objective of the new procedure is to identify heterozygous sites which require at least one additional haplotype given a set of previously chosen genotypes. For each genotype g not in the clique, if the genotype has a heterozygous site and all compatible genotypes have the same value at that site (either 0 or 1), then g is guaranteed to require one additional haplotype to be explained. Hence the lower bound can be increased by 1.

Therefore, the lower bound procedure provides a list of genotypes with an indication of the contribution of each genotype to the lower bound. Each genotype either contributes with $+2$, indicating that 2 new haplotypes will be required for explaining this genotype, or with $+1$, indicating that 1 new haplotype will be required.

This technique has been included in the PedRPoly model. In practice, the implementation of lower bounds allows the variables u associated with haplotypes affected by the lower bound to be fixed and, consequently, the clauses used for constraining the value *need not* to be generated. Indeed, if g_i is a genotype contributing with $+2$ to the lower bound, then $u_i^a = 1$ and $u_i^b = 1$. Moreover, if g_i is a genotype contributing with $+1$ to the lower bound, then either u_i^a or u_i^b can be assigned 1. The new model with integration of lower bounds will be named *PedRPoly-LB*.

4.2   Sorting Genotypes

The order in which the genotypes are organized, before the model is generated, can have an important impact on the efficiency of the solver. In particular, note that variables u designate whether a haplotype associated with a genotype is different from all previous haplotypes. We used as an heuristic the lexicographic order on the genotypes, defined by a total order on the genotype sites where $0 < 1 < 2 < ?$, i.e.

$$g_{ij} < g_{lj} \wedge (\forall_{\{k:\ k<j\}}\ g_{ik} = g_{lk}) \Rightarrow i < l. \tag{2}$$

The new model, which integrates lower bounds and where the genotypes are sorted according to the lexicographic order is named *PedRPoly-LB-Ord*.

4.3   Symmetries

Symmetry breaking is a well-known technique for pruning the search space and, therefore, contributing to the efficiency of a model. Note that, in general, in the haplotype inference problem, if a genotype g is explained by haplotype pair (h^a, h^b), then g is also explained by haplotype pair (h^b, h^a). Within pedigrees, this symmetry on pairs of haplotypes does not exist for every individual. For non-founders, symmetry is already broken by imposing that the first haplotype comes from the father and the second haplotype comes from the mother. Nonetheless, the symmetry can be broken on founders. This symmetry is broken by introducing a new constraint for each heterozygous founder, imposing that the first heterozygous site g_{ij} is explained with $h^a_{ij} = 1$ and $h^b_{ij} = 0$, i.e.

$$g_{ij} = 2 \wedge (\forall_{\{k:\ k<j\}}\ g_{ik} \neq 2) \Rightarrow t_{ij} = 1. \tag{3}$$

The new model which includes breaking symmetry on founders is named *PedRPoly-LB-Ord-Sym*.

5   Experimental Evaluation

This section has a threefold purpose. First, it illustrates the contribution of each technique described in Section 4 to the efficiency of PedRPoly. Second, it presents the results obtained using a number of constraint optimization solvers to solve the PedRPoly model, enabling the user not only to choose the best constraint solver for PedRPoly, but also indicating which solvers are more appropriate for solving Boolean constraint problems with multiple cost functions. Finally, it tests the accuracy of the PedRPoly model against the accuracy of the PedPhase approach.

Experimental Data. The experimental data was simulated using the SimPed software [17]. SimPed generates haplotypes for families, given the pedigree structure, as well as the haplotypes and their frequencies for founders. The haplotypes for founders and their frequencies were obtained from 7 real data sets of experimentally identified haplotypes [2, 29], and correspond to the A-G data sets already used in other haplotyping studies [6]. The number of SNPs range from 5 to 47. Note that haplotyping regions with tens of SNPs are still relevant in several association studies. Moreover, larger regions can always be partitioned into small blocks [36].

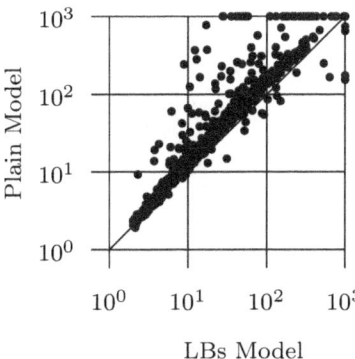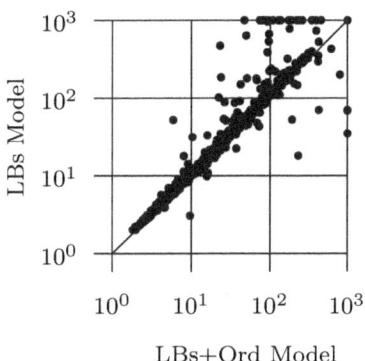

Fig. 2. CPU time comparison between models: plain PedRPoly model vs PedRPoly-LB model and PedRPoly-LB model vs PedRPoly-LB-Ord model

In addition, the same three pedigree structures used by PedPhase [20] were considered: pedigree 1 with 15 individuals, pedigree 2 with 29 individuals and pedigree 3 with 17 individuals. Pedigree 3 contains a mating loop, which means that two mating individuals have a common ancestor in the pedigree. Each simulated instance consists of 10 replicates of the given pedigree, simulating 10 different families from the same population. Hence, the number of genotypes per instance may be 150, 290 or 170. Recombination events are uniformly distributed between alleles with probabilities 0.1%, 0.5% and 1%. Three variations on missing rates were considered: 1%, 10% and 20%. For each combination of parameters, 5 independent replicates were selected, resulting in a total of 945 ($= 7 \times 3^3 \times 5$) input trials.

Genotyping errors have not been simulated. Nonetheless, genotype errors do not represent a significant limitation because they can be minimized by previously applying an appropriate error detection software [32].

Experimental Setup. All results were obtained on a Intel Xeon 5160 server (3.0GHz, 1333Mhz, 4GB) running Red Hat Enterprise Linux WS4. PedPhase ILP was run on Windows because this software is not available for Linux. Results are presented for a timeout of 1000 seconds and a memory limit of 3.5 GB.

5.1   Efficiency

This section studies the contribution of each modeling technique to improving the efficiency of PedRPoly. Furthermore, a significant number of constraint optimization solvers is tested. The use of an appropriate optimization solver with the model contributes for an efficient haplotype inference solver. In what follows, we used PedRPoly with the Boolean multilevel optimization (BMO) Max-SAT solver provided by the authors [4].

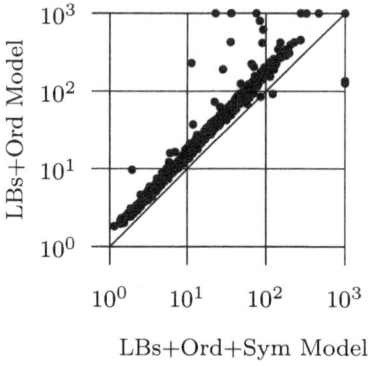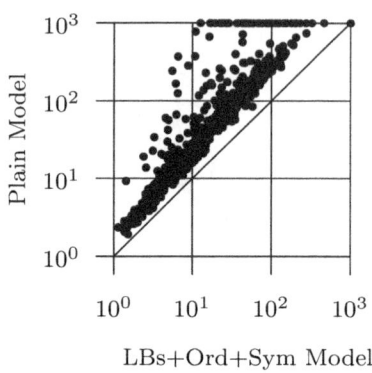

Fig. 3. CPU time comparison between models: PedRPoly-LB-Ord model vs PedRPoly-LB-Ord-Sym model and plain PedRPoly model vs PedRPoly-LB-Ord-Sym model

Lower Bounds. Figure 2 (left) provides a scatter plot which compares the performance of the plain PedRPoly with PedRPoly implementing the identification of lower bounds, within a timeout of 1000 seconds. Each point in the plot corresponds to a problem instance, where the x-axis corresponds to the CPU time required by PedRPoly-LB and the y-axis corresponds to the CPU time required by the plain PedRPoly. Points in the 10^3 lines represent instances which cannot be solved within 1000 seconds.

PedRPoly-LB reduces in half the number of instances aborted by the plain PedRPoly. The plain model aborts 59 instances while PedRPoly-LB is not able to solve 26 instances. PedRPoly-LB solves 37 instances which the plain PedRPoly aborts, although being able to solve 4 instances which PedRPoly-LB aborts. Moreover, PedRPoly-LB is faster than plain PedRPoly for more than 94% of the problem instances.

Sorting Genotypes. Figure 2 (right) compares the performance of PedRPoly-LB with PedRPoly-LB-Ord. PedRPoly-LB-Ord does not solve 12 instances but is able to solve 17 instances which PedRPoly-LB aborts. However, there are 3 instances which PedRPoly-LB solves and PedRPoly-LB-Ord is not able to solve. Moreover, PedRPoly-LB-Ord is faster than PedRPoly-LB for 86% of the instances.

Symmetries. Figure 3 (left) compares the performance of PedRPoly-LB-Ord with PedRPoly-LB-Ord-Sym. The final model is able to solve 938 out of 945 instances. PedRPoly-LB-Ord-Sym solves 7 instances which PedRPoly-LB-Ord aborts and aborts 2 instances which PedRPoly-LB-Ord solves. Moreover, PedRPoly-LB-Ord-Sym is faster than PedRPoly-LB-Ord for 99% of the instances.

Moreover, figure 3 (right) compares the performance of plain PedRPoly and PedRPoly-LB-Ord-Sym. The later is faster than the former for *all* instances,

Table 4. The PedRPoly model: comparison between models (timeout 1000 sec; memory limit 3.5 GB)

Solver	# Solved inst.	% Solved inst.	Avg run time (sec)
PedRPoly	886/945	93.76%	62.50
PedRPoly-LB	919/945	97.25%	47.30
PedRPoly-LB-Ord	933/945	98.73%	41.64
PedRPoly-LB-Ord-Sym	938/945	99.26%	24.08

and solves 52 instances which the plain model aborts. These facts illustrate the importance of the improved model in the efficiency of PedRPoly.

Table 4 summarizes the improvement achieved by combining modeling techniques. Overall, PedRPoly-LB-Ord-Sym outperforms all other models, being capable of solving 99.26% of the instances within 1000 seconds, and using an average run time of 24 seconds. In the remainder of the paper, PedRPoly-LB-Ord-Sym will be denoted simply by PedRPoly.

Solvers. A key issue for the efficiency of the haplotype inference solver is to select an adequate underlying optimization solver. In this subsection, 8 different optimization solvers were tested for solving the final version of the Ped-RPoly model. Integer linear programming, pseudo-Boolean optimization and also weighted Max-SAT solvers were considered. SCIP [1] (version 1.2.0) combines constraint programming and mixed integer programming methodologies. CPLEX (version 12.1) is an IBM/ILOG commercial linear programming optimization tool. Weighted Max-SAT solvers were also tested: MAXSAT_BMO [4], WPM1 [3], WMAXSATZ [18] (version 2.5), and INCWMAXSATZ [23]. MINISAT+ [7] and BSOLO [27] (version 3.5) are pseudo-Boolean optimization solvers, also known as 0-1 ILP solvers.

Table 5 summarizes the performance of the different solvers. Clearly, the solver which is able to solve a larger number of instances is MAXSAT_BMO, which solves 99.26% of the instances.

The second best performing solver is WPM1 which solves 96.40% of the instances. The third and fourth best performing solvers are the integer programming solvers. CPLEX solves 58.52% and SCIP solves 48.15% of the instances, followed by MINISAT+ which solves 27.51% and INCWMAXSATZ which solves 23.39% of the problem instances. BSOLO solves 16.93% and WMAXSATZ solves 1.69% of the instances. Most of the instances aborted by INCWMAXSATZ and WMAXSATZ were due to limitations in the internal data structures used by these solvers.

5.2 Accuracy

This section analyzes the gains in accuracy of PedRPoly, that integrates both HIPP with MRHC, with PedPhase [20], which uses only the MRHC approach. Two different commonly used error rates were considered. The *switch error rate*

Table 5. The PedRPoly model: comparison using different solvers (timeout 1000 sec; memory limit 3.5 GB)

Solver	# Solved inst.	% Solved inst.	Avg run time (sec)
MaxSat_bmo	938/945	99.26%	24.08
WPM1	911/945	96.40%	14.75
Cplex	553/945	58.52%	84.73
Scip	455/945	48.15%	139.07
MiniSat+	260/945	27.51%	238.45
IncWMaxSatz	221/945	23.39%	71.04
Bsolo	160/945	16.93%	170.26
WMaxSatz	16/945	1.69%	113.16

measures the percentage of possible switches in haplotype orientation, used to recover the correct phase in an individual [24]. Missing alleles are not considered for computing the switch error. The *missing error rate* (or *genotype inference error rate*) is the percentage of incorrectly inferred missing data [28].

Note that instances for which at least one of the solvers is unable to give a solution have been removed from the comparison. PedPhase is able to solve 99.8% of the instances, whereas and PedRPoly is able to solve 99.3%. As a result, 9 out of 945 instances have been left out.

Figure 4 presents a bar graph comparing the switch error rate of PedRPoly with the switch error rate of PedPhase. Results have been organized by parameter value: missing rate, recombination rate and pedigree. Each value is the average of the error rate for the instances generated with the corresponding parameter value. PedRPoly is more accurate than PedPhase for 67.09% of the instances. The two solvers have equal error rates for 19.55% of the instances. For 13.35% of the instances, PedRPoly is less accurate than PedPhase.

Figure 5 presents a bar graph for evaluating the missing error rate of the two tools. PedRPoly is more accurate than PedPhase for 73.93% of the instances. The two solvers have equal error rate for 12.39% of the instances. For 13.68% of the instances, PedRPoly is less accurate than PedPhase. Indeed, the population information included by the PedRPoly model is shown to be particularly important for inferring missing genotypes. Overall, we can conclude that PedRPoly consistently outperforms PedPhase in terms of accuracy.

Although the goals of the paper are to show that MRHC combined with HIPP has better accuracy than MRHC alone, and the development of optimizations to the plain model, two distinct statistical methods for haplotype inference within pedigrees were also evaluated. The methods evaluated are Superlink [8] and PhyloPed [15], and both exhibit error rates higher than PedRPoly.

Finally, the number of distinct haplotypes in the PedRPoly solution and in the PedPhase solution was compared with the number of haplotypes used in the real solution. The number of haplotypes in the PedRPoly solution is exactly the same as in the real solution for 60% of the instances, and for more than 99.6% the number of haplotypes in the PedRPoly solution differs from the number of haplotypes in the real solution by less than 5 haplotypes. PedPhase solutions,

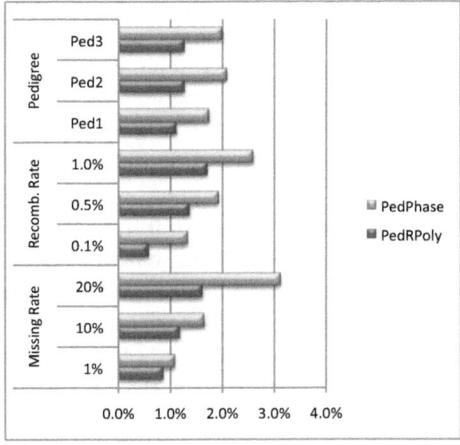

Fig. 4. Switch error: comparing PedRPoly and PedPhase

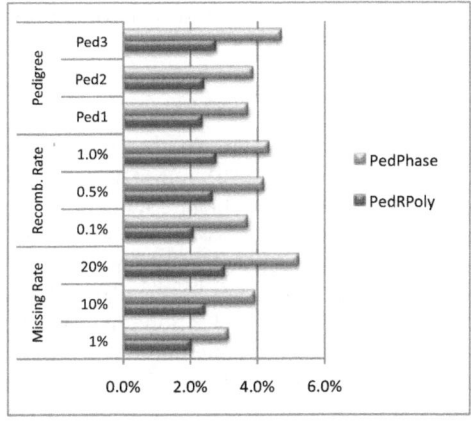

Fig. 5. Missing error: comparing PedRPoly and PedPhase

however, are less similar to the real solutions with respect to the number of haplotypes. For the same set of instances, PedPhase has the same number of haplotypes as the real solution for 12.3% of the instances, and differs from the real solution by less than 5 haplotypes for 45.8% of the instances.

6   Conclusions

This paper addresses the problem of haplotype inference from pedigrees, and proposes a new Boolean optimization model for haplotype inference which combines the pure parsimony approach with the minimum recombinant approach.

Moroever, the paper details the integration of well-known modeling techniques exploited in order to improve the performance of the method. These techniques include integration of lower bounds, ordering heuristics and symmetry breaking, as well as the selection of an appropriate constraint solver. The new PedRPoly approach was tested on a set of instances of considerable dimension. Experimental results show that the new approach is both accurate and efficient when compared to other methods.

The problem instances generated by the combined model represent challenging combinatorial optimization problems, related with multi-objective optimization. A number of techniques was suggested to improve the performance for this class of problem instances. Future research will address further optimizations to the model, aiming at improving both accuracy and efficiency. The topic of improved efficiency will also involve the development of solvers capable of solving Boolean-based multi-objective optimization problems.

Acknowledgments. This work is partially supported by Fundação para a Ciência e Tecnologia under research project SHIPs (PTDC/EIA/64164/2006) and PhD grant (SFRH/BD/28599/2006), and by Microsoft under contract 2007-017 of the Microsoft Research PhD Scholarship Programme.

References

[1] Achterberg, T., Berthold, T., Koch, T., Wolter, K.: Constraint Integer Programming: A New Approach to Integrate CP and MIP. In: Trick, M.A. (ed.) CPAIOR 2008. LNCS, vol. 5015, pp. 6–20. Springer, Heidelberg (2008)

[2] Andrés, A., Clark, A., Shimmin, L., Boerwinkle, E., Sing, C., Hixson, J.: Understanding the accuracy of statistical haplotype inference with sequence data of known phase. Genetic Epidemiology 31(7), 659–671 (2007)

[3] Ansótegui, C., Bonet, M.L., Levy, J.: Solving (Weighted) Partial MaxSAT through Satisfiability Testing. In: Kullmann, O. (ed.) SAT 2009. LNCS, vol. 5584, pp. 427–440. Springer, Heidelberg (2009)

[4] Argelich, J., Lynce, I., Marques-Silva, J.: On solving Boolean multilevel optimization problems. In: International Joint Conference on Artificial Intelligence (IJCAI 2009), pp. 393–398 (2009)

[5] Cheng, I., Penney, K.L., Stram, D.O., Le Marchand, L., Giorgi, E., Haiman, C.A., Kolonel, L.N., Pike, M., Hirschhorn, J., Henderson, B.E., Freedman, M.L.: Haplotype-based association studies of IGFBP1 and IGFBP3 with prostate and breast cancer risk: the multiethnic cohort. Cancer Epidemiol Biomarkers Prev. 15(10), 1993–1997 (2006)

[6] Climer, S., Jäger, G., Templeton, A.R., Zhang, W.: How frugal is mother nature with haplotypes? Bioinformatics 25(1), 68–74 (2009)

[7] Eén, N., Sörensson, N.: Translating pseudo-Boolean constraints into SAT. Journal on Satisfiability, Boolean Modeling and Computation 2, 1–26 (2006)

[8] Fishelson, M., Dovgolevsky, N., Geiger, D.: Maximum likelihood haplotyping for general pedigrees. Human Heredity 59(1), 41–60 (2005)

[9] Graça, A., Lynce, I., Marques-Silva, J., Oliveira, A.: Haplotype inference combining pedigrees and unrelated individuals. In: Workshop on Constraint Based Methods for Bioinformatics (WCB 2009), pp. 27–36 (2009)

[10] Graça, A., Marques-Silva, J., Lynce, I., Oliveira, A.L.: Efficient Haplotype Inference with Pseudo-boolean Optimization. In: Anai, H., Horimoto, K., Kutsia, T. (eds.) AB 2007. LNCS, vol. 4545, pp. 125–139. Springer, Heidelberg (2007)

[11] Graça, A., Marques-Silva, J., Lynce, I., Oliveira, A.L.: Efficient Haplotype Inference with Combined CP and OR Techniques. In: Trick, M.A. (ed.) CPAIOR 2008. LNCS, vol. 5015, pp. 308–312. Springer, Heidelberg (2008)

[12] Gusfield, D.: Haplotype Inference by Pure Parsimony. In: Baeza-Yates, R., Chávez, E., Crochemore, M. (eds.) CPM 2003. LNCS, vol. 2676, pp. 144–155. Springer, Heidelberg (2003)

[13] Haines, J.L.: Chromlook: an interactive program for error detection and mapping in reference linkage data. Genomics 14(2), 517–519 (1992)

[14] Kimura, M.: The number of heterozygous nucleotide sites maintained in a finite population due to steady flux of mutations. Genetics 61(4) (1969)

[15] Kirkpatrick, B., Rosa, J., Halperin, E., Karp, R.M.: Haplotype Inference in Complex Pedigrees. In: Batzoglou, S. (ed.) RECOMB 2009. LNCS, vol. 5541, pp. 108–120. Springer, Heidelberg (2009)

[16] Lancia, G., Pinotti, C.M., Rizzi, R.: Haplotyping populations by pure parsimony: complexity of exact and approximation algorithms. INFORMS Journal on Computing 16(4), 348–359 (2004)

[17] Leal, S.M., Yan, K., Müller-Myhsok, B.: SimPed: A simulation program to generate haplotype and genotype data for pedigree structures. Human Heredity 60(2), 119–122 (2005)

[18] Li, C.M., Manyà, F., Mohamedou, N., Planes, J.: Exploiting Cycle Structures in Max-SAT. In: Kullmann, O. (ed.) SAT 2009. LNCS, vol. 5584, pp. 467–480. Springer, Heidelberg (2009)

[19] Li, J., Jiang, T.: Efficient inference of haplotypes from genotypes on a pedigree. Journal of Bioinformatics and Computational Biology 1(1), 41–69 (2003)

[20] Li, J., Jiang, T.: Computing the minimum recombinant haplotype configuration from incomplete genotype data on a pedigree by integer linear programming. Journal of Computational Biology 12(6), 719–739 (2005)

[21] Li, X., Li, J.: Comparison of haplotyping methods using families and unrelated individuals on simulated rheumatoid arthritis data. In: BMC Proceedings, pp. S1–S55 (2007)

[22] Li, X., Li, J.: Efficient haplotype inference from pedigree with missing data using linear systems with disjoint-set data structures. In: International Conference on Computational Systems Bioinformatics (CSB 2008), pp. 297–307 (2008)

[23] Lin, H., Su, K., Li, C.M.: Within-problem learning for efficient lower bound computation in Max-SAT solving. In: National Conference on Artificial Intelligence (AAAI 2008), pp. 351–356 (2008)

[24] Lin, S., Chakravarti, A., Cutler, D.J.: Haplotype and missing data inference in nuclear families. Genome Research 14(8), 1624–1632 (2004)

[25] Liu, L., Xi, C., Xiao, J., Jiang, T.: Complexity and approximation of the minimum recombinant haplotype configuration problem. Theoretical Computer Science 378(3), 316–330 (2007)

[26] Lynce, I., Marques-Silva, J., Prestwich, S.: Boosting haplotype inference with local search. Constraints 13(1), 155–179 (2008)

[27] Manquinho, V., Marques-Silva, J.: Effective lower bounding techniques for pseudo-Boolean optimization. In: Design, Automation and Test in Europe Conference and Exhibition (DATE 2005), pp. 660–665 (2005)

[28] Marchini, J., Cutler, D., Patterson, N., Stephens, M., Eskin, E., Halperin, E., Lin, S., Qin, Z.S., Munro, H.M., Abecassis, G.R., Donnelly, P., International HapMap Consortium: A comparison of phasing algorithms for trios and unrelated individuals. American Journal of Human Genetics 78(3), 437–450 (2006)

[29] Orzack, S.H., Gusfield, D., Olson, J., Nesbitt, S., Subrahmanyan, L., Stanton, V.P.: Analysis and exploration of the use of rule-based algorithms and consensus methods for the inferral of haplotypes. Genetics 165(2), 915–928 (2003)

[30] Pei, Y., Zhang, L., Li, J., Papasian, C.J., Deng, H.-W.: Analyses and comparison of accuracy of different genotype imputation methods. PLoS ONE 3(10) (2008)

[31] Qian, D., Beckmann, L.: Minimum-recombinant haplotyping in pedigrees. American Journal of Human Genetics 70(6), 1434–1445 (2002)

[32] Sánchez, M., Givry, S., Schiex, T.: Mendelian error detection in complex pedigrees using weighted constraint satisfaction techniques. Constraints 13(1-2), 130–154 (2008)

[33] The International HapMap Consortium: A second generation human haplotype map of over 3.1 million SNPs. Nature 449, 851–861 (2007)

[34] Wang, L., Xu, Y.: Haplotype inference by maximum parsimony. Bioinformatics 19(14), 1773–1780 (2003)

[35] Wijsman, E.M.: A deductive method of haplotype analysis in pedigrees. American Journal of Human Genetics 41(3), 356–373 (1987)

[36] Zhang, K., Qin, Z., Chen, T., Liu, J.S., Waterman, M.S., Sun, F.: HapBlock: haplotype block partitioning and tag SNP selection software using a set of dynamic programming algorithms. Bioinformatics 21(1), 131–134 (2005)

[37] Zhang, K., Sun, F., Zhao, H.: HAPLORE: a program for haplotype reconstruction in general pedigrees without recombination. Bioinformatics 21(1), 90–103 (2005)

MABSys: Modeling and Analysis
of Biological Systems

François Lemaire and Asli Ürgüplü

University of Lille I, LIFL
Villeneuve d'Ascq, France
{Francois.Lemaire,Asli.Urguplu}@lifl.fr

Abstract. We present the MABSys package which gathers, as much as possible, some functions to carry out the modeling of biochemical reaction networks, their qualitative analysis and the exact simplification of systems of ordinary differential equations. The main functions are illustrated with examples including the corresponding commands. Then we discuss Tyson's negative feedback oscillator model and the parameters values for which this system oscillates.

Keywords: software design, systems of ordinary differential equations, qualitative analysis.

1 Introduction

This article presents the MABSys package, a pilot implementation conceived and developed by the authors using the Maple computer algebra software. MABSys assists biologists and bioinformaticians in the analysis of their biological systems modeled by medium size (about twenty coordinates) systems of ordinary differential equations (ODEs). Such mathematical models (assumed to be continuous dynamical systems) are of precious help to understand biological systems. They describe qualitative dynamics of a model where the intuition is no more sufficing. Furthermore, we have chosen to be accessible to non-expert users in mathematics: the use of our algorithms, that are based on the algebraic elimination, Lie symmetries etc., does not require any mathematical knowledge.

MABSys includes three main parts: the modeling of biochemical reaction networks by means of ODEs, the exact symbolic simplification of these dynamical systems and the qualitative analysis tools to retrieve informations about the behavior of the model. The main structure of MABSys is given in figure 1. Every rectangular box corresponds to a set of functions and every ellipsis to an input or an output element. The arrows indicate directions of function sequences.

Section 2 presents the related works about existing softwares on systems biology and similar domains. Section 3 introduces the main structure of MABSys with illustrations on simple examples. Section 4 treats Tyson's negative feedback oscillator using MABSys and discusses the results. Section 5 helps with the installation of MABSys and last section concludes the paper.

K. Horimoto, M. Nakatsu, and N. Popov (Eds.): ANB 2011, LNCS 6479, pp. 57–75, 2012.

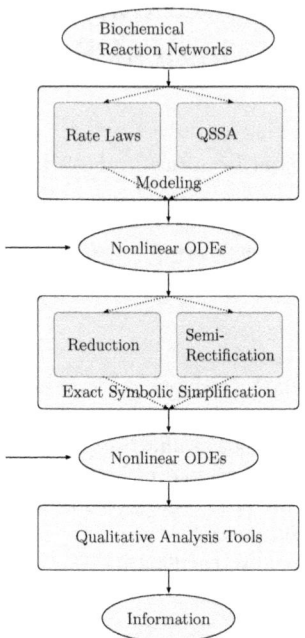

Fig. 1. Main structure of MABSys

2 Related Works

There exist many approaches used for the comprehension of biological systems (for example see [23,29]). An exhaustive list of available softwares can be found in SBML Software Guide (see [15]). Many of these tools are based on simulations, not symbolic analysis because of the complexity of the studied models. For instance, COPASI is a software application for simulation and analysis of biochemical networks and GNA is a computer tool for modeling and simulation of genetic regulatory networks in the form of piecewise-linear differential equations. Our MABSys package treats systems of ODEs with symbolic approaches.

Computer algebra softwares need a formal structuring of biological systems. Diagram editors for drawing such systems as CellDesigner or JDesigner permit to export networks in the form of SBML. For now, the description made in MABSys includes only the most important parts of biochemical reactions in their simplest way which is sufficient for their modeling by means of ODEs. In the future, it could be interesting to integrate SBML format.

There exist also widely used general strategies for simplification of differential systems. The lumping, the sensitivity analysis or the time-scale analysis (see [24]), all decrease the number of coordinates but they cause a loss of information about individual original coordinates. In our work, we keep the explicit relationships between the original and the simplified systems. Our algorithms are based on the well-known Lie symmetry theory with new approaches.

It generalizes for example the dimensional analysis (see [16]) which is a classical reduction method based on the units of coordinates. The existing softwares on this domain (see ch. 17 of [11], [13,21]) are not suitable for non-experts in mathematics. In addition, they do not guarantee the polynomial complexity of the algorithms contrarily to the simplification methods used in **MABSys**.

3   Main Structure of **MABSys**

MABSys is a package which is oriented towards biochemical reaction networks and their analysis. This section details its functionalities. We illustrate the use of main functions but for more specific aspects, options and auxiliary functions, we refer the reader to the associated help pages [20].

3.1   Description of Biochemical Reaction Networks

Functions: `GetFastReactions`, `GetReactionRate`, `GetReactionReactants`, `GetReactionProducts`, `GetSlowReactions`, `IsReactionFast`, `NewReaction`

Types: `Reaction`, `ReactionSystem`

In **MABSys**, a biochemical reaction network describes interactions between macromolecules (genes, mRNAs, proteins) towards some process as binding, release, synthesis, degradation, transformation and signal responses. Such networks are denoted by the usual chemical notation which is used to generate models of ODEs. For instance, the reaction $A + B \rightarrow C$ between 3 chemical species is represented by its reactants (A,B), its product (C) and the rate law of the reaction. Every one-way reaction is represented by a simple `Reaction` data structure. It is created by the `NewReaction` constructor and needs: the reactants and the products given as a linear combination of species names, the rate law of the system. A classical way of modeling reactions is to use the mass-action law indicated by the keyword `MassActionLaw` but one can also choose another reaction law in the form of rational fractions (for instance Hill functions) indicated by the keyword `CustomizedLaw`. For the mass-action law, one must give the associated rate constant and for a customized law the whole rate. An optional boolean `fast` that indicates the velocity of the reaction can be specified. The `true` value is used for fast reactions and the `false` value is used for slow reactions. The default value is false. This option is important in order to apply the QSSA (see § 3.2) on the reaction system. A system of reaction that corresponds to the `ReactionSystem` data structure is a list of reactions.

Remark 1. Because the **MABSys** package is developed in `Maple`, in this paper, piece of codes are detailed by respecting the `Maple` syntax. Each line in the code frames that begins with the symbol ">" stands for the commands sent to `Maple`. Each line without this symbol stands for an output. The symbol "#" indicates a comment. Also, the keywords that belong to **MABSys** are written in bold so that one can distinguish them from standard `Maple` commands.

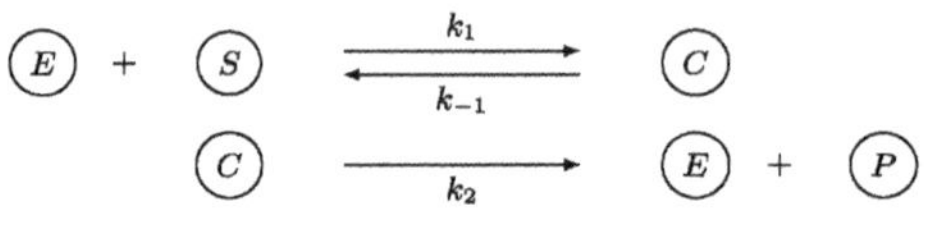

Fig. 2. Basic enzymatic biochemical reaction system

Example 1. Let us see a basic enzymatic reaction system given in figure 2 and its construction in **MABSys**. This system describes the transformation of a substrate S into a product P under the action of the enzyme E. Meanwhile, an intermediate complex C is produced. Each reaction follows the mass-action law and the associated rate constants are indicated above or below of the arrows. The system contains 3 one-way reactions defined in two timescales. The first two reactions are considered fast w.r.t. the third one meaning that their rate constants k_1, k_{-1} are supposed to be greater than the third one k_2.

```
> # The biochemical reactions
> R1 := NewReaction(E+S,C,MassActionLaw(k1),fast=true);
        R1 := Reaction([E, S], [C], MassActionLaw(k1), true)

> R2 := NewReaction(C,E+S,MassActionLaw(km1),fast=true);
        R2 := Reaction([C], [E, S], MassActionLaw(km1), true)

> R3 := NewReaction(C,E+P,MassActionLaw(k2));
        R3 := Reaction([C], [E, P], MassActionLaw(k2), false)

> type(R1, Reaction);
        true

> # The biochemical reaction system
> RS := [R1,R2,R3];
        RS := [Reaction([E, S], [C], MassActionLaw(k1), true),
               Reaction([C], [E, S], MassActionLaw(km1), true),
               Reaction([C], [E, P], MassActionLaw(k2), false)]

> type(RS, ReactionSystem);
        true

> # Auxiliary functions
> IsReactionFast(R1);
        true

> GetFastReactions(RS);
        [Reaction([E, S], [C], MassActionLaw(k1), true),
         Reaction([C], [E, S], MassActionLaw(km1), true)]

> GetSlowReactions(RS);
        [Reaction([C], [E, P], MassActionLaw(k2), false)]
```

As you see, in **MABSys** there are also many auxiliary functions to manipulate the reactions such as to extract the name of products, the name of reactants, to get fast or slow reactions, to get rate constants etc. For all these functions see the associated help pages [20].

Mass-action Law. The mass-action law is a classical way of defining dynamics of a reaction in function of the concentration of the present species and the rate constant which quantifies its speed. Mainly, this law tells that, for an elementary reaction, the reaction rate is proportional to the reactant concentrations raised to a particular power which is their stoichiometric coefficient and to the rate constant.

Example 2. In the following example we create a reaction where two molecules of A and one molecule of B are transformed into the product C by following the mass-action law with k as rate constant. By default, the reaction is considered as slow. Then we call some MABSys commands to illustrate the associated functionalities.

```
> # Mass-action law
> R4 := NewReaction(2*A+B,C,MassActionLaw(k));
          R4 := Reaction([2 A, B], [C], MassActionLaw(k), false)

> GetReactionReactants(R4);
          [2 A, B]

> GetReactionProducts(R4);
          [C]

> GetReactionRate(R4);
             2
          k A  B
```

Customized Law. A customized law lets the user to indicate the rate law of a reaction in rational function. Again, the names of species indicate associated concentrations and the new letters are considered as parameters. This option is important, for example, in order to manage the signal effects on the biochemical reactions.

Example 3. In this code example we create two reactions. The transformation consists of converting a protein P into another protein Q by respecting the given customized law. We also create the synthesis of the protein P without considering the source and assuming that the reaction follows a customized law in the form of a Hill function depending on a gene G. Both are considered as slow.

```
> # Customized law
> R5 := NewReaction(P,Q,CustomizedLaw(k*P/(V+P)));
                                        k P
          R5 := Reaction([P], [Q], CustomizedLaw(-----), false)
                                        V + P

> GetReactionRate(R5);
          k P
          -----
          V + P

> R6 := NewReaction(0,P,CustomizedLaw(G/(G+theta)));
                                          G
          R6 := Reaction([], [P], CustomizedLaw(---------), false)
                                        G + theta

> GetReactionRate(R6);
             G
          ---------
          G + theta
```

3.2   Modeling by Means of ODEs

Two intimately related kinds of modeling by means of ODEs are available for these networks. On the one hand, one can use directly the rate laws of the reactions to construct the so called *basic model.* On the other hand, the new algorithm that performs the quasi-steady state approximation (see [4]) assures a modeling by reducing this basic model. The quasi-steady state approximation is preferable if the network possesses two timescales, fast and slow reactions, since it can lead to a simpler model than the basic one.

Basic Modeling

Functions: `RateVector, ReactionSystem2ODEs, StoichiometricMatrix`

One of the simplest classical way of modeling a biological system consists of constructing a model by following directly the rate laws (mass-action or customized) of the reactions. The basic modeling of a biological system of r one-way reactions that involves k species denoted by X by means of ODEs requires the associated rate vector V of dimension r and the stoichiometric matrix M of dimension $k \times r$. In vector-valued notations, the basic model is given by the formula $\dot{X} = M V$. This model can be obtained by the `ReactionSystem2ODEs` function of `MABSys`.

Example 4. The computation of the basic model that corresponds to the enzymatic reaction system of figure 2 follows.

$$
\begin{pmatrix} \dot{E} \\ \dot{S} \\ \dot{C} \\ \dot{P} \end{pmatrix} = \begin{pmatrix} -1 & 1 & 1 \\ -1 & 1 & 0 \\ 1 & -1 & -1 \\ 0 & 0 & 1 \end{pmatrix} \begin{pmatrix} k_1\,E\,S \\ k_{-1}\,C \\ k_2\,C \end{pmatrix} \quad \Leftrightarrow \quad \begin{cases} \dot{E} = -k_1\,E\,S + k_{-1}\,C + k_2\,C, \\ \dot{S} = -k_1\,E\,S + k_{-1}\,C, \\ \dot{C} = k_1\,E\,S - k_{-1}\,C - k_2\,C, \\ \dot{P} = k_2\,C. \end{cases} \tag{1}
$$

This biological system is already encoded in the variable RS in example 1. Here are the associated functions outputs.

```
> # The rate vector
> RateVector(RS);
        [k1 E S]
        [       ]
        [km1 C  ]
        [       ]
        [ k2 C  ]

> # The stoichiometric matrix
> StoichiometricMatrix(RS, [E,S,C,P]);
        [-1    1    1]
        [           ]
        [-1    1    0]
        [           ]
        [ 1   -1   -1]
        [           ]
        [ 0    0    1]

> # Basic modeling
> ReactionSystem2ODEs(RS, [E,S,C,P]);
        d
        [-- E(t) = -k1 E(t) S(t) + km1 C(t) + k2 C(t),
        dt
        d
        -- S(t) = -k1 E(t) S(t) + km1 C(t),
        dt
        d
        -- C(t) = k1 E(t) S(t) - km1 C(t) - k2 C(t),
        dt
        d
        -- P(t) = k2 C(t)]
        dt
```

Quasi Steady-State Approximation

Function: `ModelReduce`

There are different ways to perform the quasi-steady state approximation (QSSA) which is an inexact simplification method. The classical one which consists to replace some variables with their steady state values is useful in many cases. However, for example when one is interested in the timescales over which the system equilibrates or in the period and the amplitude of oscillations for an oscillating system, this classical QSSA must be adjusted. Indeed, the separation of timescales is the key for observing nontrivial behaviors. In biology, many processes like dimerization occur faster than others. In [4], the QSSA method considered by [32] is reformulated and made fully algorithmic. In MABSys the function ModelReduce implements this algorithm. It attempts to construct a reduced model that represents biochemical reaction networks behavior with less chemical species thus a simpler model than the basic one. The algorithm can be expressed by means of either differential elimination methods (see [3,2,25]) or regular chains using [18]. In [1], authors apply the QSSA to a family of networks proposed in [5] and obtain a more precise model.

For biological systems, two classes of reactions are considered: slow and fast ones. The idea is to study the dynamics of slow reactions, assuming that the fast ones are at quasi-equilibrium, thereby removing from the system of ODEs, the differential equations which describe the evolution of the variables at quasi-equilibrium.

Example 5. The QSSA applied on the basic enzymatic reaction system of figure 2 yields a formula that describes the dynamics of S automatically:

$$\frac{\mathrm{d}S}{\mathrm{d}t} = -\frac{V_m\, S\,(K+S)}{K\, E_0 + (K+S)^2} \tag{2}$$

where $V_m = k_2\, E_0$ and K are parameters and E_0 is the initial concentration of E. Moreover, the classical formula given in the early xx[th] century for the same problem (see [12,22,6]) rely on a few extra assumptions than QSSA. Our result seems more accurate when S is not supposed to be greater than E_0. For numerical simulations that verify these phenomena see [4,1]. Here follows the output of the function ModelReduce executed on the variable RS, the system reaction constructed within the example 1.

```
> # Modeling by QSSA
> QSSAModel := ModelReduce(RS, [E,C,P,S], useConservationLaws=true)[1,1]:
> # The further simplifications
> QSSAModel := simplify(subs({k1=km1/K,C_0=0,P_0=0,k2=Vm/E_0},QSSAModel)):
> # The last equation correspond to the substrate.
> QSSAModelS := QSSAModel[-1];
              d                       Vm S(t) (K + S(t))
    QSSAModelS := -- S(t) = - ----------------------------
              dt                        2                2
                              2 S(t) K + S(t)  + E_0 K + K
```

3.3   Exact Symbolic Simplifications

Any system of polynomial ODEs, coming from any scientific context, can be treated by the exact simplifications part i.e. by the reduction and the semi-rectification methods. These exact simplifications are based on the classical Lie symmetry theory which is adapted in an original way to the modeling in biology. The reduction of the parameter set of a model consists in *eliminating* some parameters thus in decreasing their number. The semi-rectification of a model consists in understanding more the influence of the parameters values on the system dynamics. The goal of these simplifications is to reorganize the model coordinates so that the resulting equivalent model can be more easily tractable.

The use of our algorithms does not require any knowledge about Lie symmetries. However, even if the simplification algorithms are fully automatic, the choice of the parameters to eliminate, the parameters and the variables to adjust, etc. remain under the control of the user.

Implementation Remarks The `MABSys` package, where these simplification algorithms are implemented, relies on the `ExpandedLiePointSymmetry` package (see [27]) for the computation of (scaling type) Lie point symmetries. The complexity of the algorithm employed for this issue is polynomial in the input size (see [26,14] and proposition 4.2.8 in [31]). This gain of complexity arises mostly from the limitation to only scalings and the restriction of the general definition of Lie symmetries.

The structure of used Lie symmetries in `MABSys` permits to get the exact relations associated to model simplification and to respect the positivity property of biological quantities (parameters and concentrations) when the system models a biological phenomenon. It is possible to generalize the methods used in `MABSys`, for example to the reduction of state variables, to the simplification of discrete dynamical systems etc. These generalizations are implemented in the `ExpandedLiePointSymmetry` package (see [27]).

Reduction

Function: `InvariantizeByScalings`

The objective of the reduction method is to eliminate some parameters by constructing new coordinates and thus to decrease their number. Thanks to our algorithm, there exists a bijection between the positive solutions of the original system of ODEs and these of the reduced one.

The arguments of the associated function `InvariantizeByScalings` are the model to reduce and two lists to guide this reduction: the list of parameters assumed positive given by decreasing order of preference to remove from the model and the list of remaining coordinates. For now, this second list must contain the state variables. The algorithm takes each parameter in the first list and tries to eliminate it before considering the next one. It may happen that some of these parameters cannot be eliminated because of the system structure. The output is composed of three objects: the reduced system of ODEs, the change of coordinates that gives the exact relations between the original and the reduced systems coordinates and the list of parameters that were eliminated.

Example 6. Let us illustrate the outputs of the reduction process on the two-species oscillator that models the dynamics of a biochemical reactions network:

$$\begin{cases} \frac{dx}{dt} = a - k_1\,x + k_2\,x^2\,y, \\ \frac{dy}{dt} = b - k_2\,x^2\,y. \end{cases} \tag{3}$$

This system has two state variables x, y and depends on 4 parameters: a, b, k_1 and k_2. Assuming that the system parameters are positive, our algorithm permits to rewrite it in a new coordinate set where the system depend on only 2 parameters instead of 4. According to the algorithms given in [31], the new coordinate set $\widehat{Z} = \left(\widehat{t}, \widehat{x}, \widehat{y}, \widehat{k}_1, \widehat{k}_2, \widehat{a}, \widehat{b}\right)$ is defined by:

$$\widehat{t} = t\,k_1, \ \widehat{x} = \frac{x\,k_1}{a}, \ \widehat{y} = \frac{y\,k_1}{a}, \ \widehat{k}_1 = k_1, \ \widehat{k}_2 = \frac{k_2\,a^2}{k_1^3}, \ \widehat{a} = a, \ \widehat{b} = \frac{b}{a}. \quad (4)$$

This change of coordinates is reversible by construction. The substitution of its inverse gives the reduced system:

$$\begin{cases} \frac{d\widehat{x}}{d\widehat{t}} = 1 - \widehat{x} + \widehat{k}_2\,\widehat{x}^2\,\widehat{y}, \\ \frac{d\widehat{y}}{d\widehat{t}} = \widehat{b} - \widehat{k}_2\,\widehat{x}^2\,\widehat{y}. \end{cases} \quad (5)$$

Remark that the parameters $\widehat{a}$ and $\widehat{k}_1$ do not appear in the system definition. However, this new system (5) is *equivalent* to the system (3) by the exact relations (4). There is a bijection between the positive solutions (coming from the positive coordinates values) of these two systems. Here follows the MABSys code that computes automatically this simplification. Remark that the output notation in the code example does not differentiate the new coordinates from the old ones for the sake of computational simplicity. The outputs must be interpreted exactly as above.

```
> # Model
> ODEs := [diff(x(t),t)=a-k1*x(t)+k2*x(t)^2*y(t), diff(y(t),t)=b-k2*x(t)^2*y(t)];
            d                          2            d                       2
   ODEs := [-- x(t) = a - k1 x(t) + k2 x(t) y(t), -- y(t) = b - k2 x(t) y(t)]
            dt                                     dt

> # Reduction
> out := InvariantizeByScalings(ODEs,[a,b,k1,k2],[x,y]):
> ReducedODEs := out[1];
                d                      2            d                  2
   ReducedODEs := [-- x(t) = 1 - x(t) + k2 x(t) y(t), -- y(t) = b - k2 x(t) y(t)]
                dt                                     dt

> CoC1 := out[2];
                           2
                        k2 a                     x k1     y k1
           CoC1 := [b = b/a, k2 = -----, t = t k1, x = ----, y = ----]
                           3                       a        a
                         k1
> EliminatedParams := out[3];
        EliminatedParams := [a, k1]
```

Semi-rectification

Function: SemiRectifySteadyPoints

The objective of the semi-rectification method is to make some parameters to appear as factors in the right-hand side of ODEs (see [19]). This property simplifies the expressions of the associated steady points and permits to distinguish the roles of the parameters. For example, some parameters matter to decide only the nature (attractor or repellor) of steady points. Other ones are involved also to define their place.

The associated function `SemiRectifySteadyPoints` needs the model to reduce and two lists to guide this semi-rectification. One must specify the list of parameters assumed positive given by decreasing order of preference for the semi-rectification and the list of remaining coordinates. For now, this second list must contain the state variables. The output includes: the semi-rectified system of ODEs, the algebraic system that defines the steady points of this new differential system, the change of coordinates that gives the exact relations between the original and the semi-rectified systems and the list of parameters that do not appear anymore in the solutions of the algebraic system that defines the steady points of the new differential system.

Example 7. Let us illustrate the outputs of the semi-rectification process on the reduced two-species oscillator given in (5). Remark that this system cannot be reduced anymore by our reduction process. Assuming that the parameters $\widehat{b}$ and $\widehat{k}_2$ are positive, our semi-rectification algorithm permits to define a new coordinate set $\widetilde{Z} = \left(\widetilde{t}, \widetilde{x}, \widetilde{y}, \widetilde{k}_1, \widetilde{k}_2, \widetilde{a}, \widetilde{b}\right)$ where $\widetilde{y} = \widehat{y}\,\widehat{k}_2$ and all other coordinates remain the same. Again, this change of coordinates is reversible by construction. The substitution of its inverse gives the semi-rectified system:

$$\begin{cases} \frac{d\widetilde{x}}{d\widetilde{t}} = 1 - \widetilde{x} + \widetilde{x}^2\,\widetilde{y}, \\ \frac{d\widetilde{y}}{d\widetilde{t}} = \left(\widetilde{b} - \widetilde{x}^2\,\widetilde{y}\right)\widetilde{k}_2. \end{cases} \tag{6}$$

Remark that the parameter $\widetilde{k}_2$ does not appear anymore in the algebraic system defining the steady points of this final system of ODEs:

$$\begin{cases} 1 - \widetilde{x} + \widetilde{x}^2\,\widetilde{y} = 0, \\ \widetilde{b} - \widetilde{x}^2\,\widetilde{y} = 0. \end{cases} \tag{7}$$

There is a bijection between the positive solutions (coming from the positive coordinates values) of (5) and (6). Here follows the `MABSys` code that computes automatically this simplification. Once more, remark that the output notation in the code example does not differentiate the new coordinates from the old ones for the sake of computational simplicity. The outputs must be interpreted exactly as above.

```
> # Semi-rectification
> ReducedODEs;
        d                    2      d                      2
      [-- x(t) = 1 - x(t) + k2 x(t)  y(t), -- y(t) = b - k2 x(t)  y(t)]
        dt                              dt
> out := SemiRectifySteadyPoints(ReducedODEs, [b,k2], [x,y]):
> out := out[1]:
> FinalODEs := out[1];
        d                    2      d                  2
    FianlODEs := [-- x(t)=1 - x(t) + x(t)  y(t), -- y(t)=(b - x(t)  y(t)) k2]
        dt                              dt

> FinalAlgSys := out[2];
                        2          2
    FinalAlgSys := [1 - x + x  y, b - x  y]

> CoC2 :=out[3];
        CoC2 := [y = y k2]

> FactorizedParams := out[4];
        FactorizedParams := [k2]
```

Remark 2. The semi-rectification process can be improved using the *triangular-ization* notion via the `RegularChains` package [18] of `Maple`. This option can help to simplify more the studied system. The disadvantage is that the complexity of the associated computations is not polynomial in the worst case because this method requires the algebraic elimination. In this paper we do not detail this option for the sake of clarity.

Change of Coordinates

Functions: `ApplyChangeOfCoord`,
  `ComposeChangeOfCoord`, `InverseChangeOfCoord`

The change of coordinates found by the exact simplification methods gives the relationships between the old coordinates $Z = (z_1, \ldots, z_n)$, in which the original system is written, and the new coordinates $\widetilde{Z} = (\widetilde{z}_1, \ldots, \widetilde{z}_n)$, in which the simplified system is written. In `MABSys`, these changes of coordinates are restricted to monomial maps of the form:

$$\widetilde{z}_j = \prod_{k=1}^{n} z_k^{C_{k,j}} \quad \forall j \in \{1, \ldots, n\} \tag{8}$$

where the $C_{k,j}$'s are elements of a $n \times n$ invertible matrix with rational coefficients (see § 2.1 in [19]). These changes of coordinates are invertible.

Example 8. In this example, we illustrate the functions that inverse changes of coordinates, apply them to a given system and compose them with each other.

```
> # The inverse of CoC1
> InverseChangeOfCoord (CoC1);
               3
            k2 k1           t           x a        y a
  [b = a b, k2 = ------, t = ----, x = ----, y = ---]
               2            k1          k1         k1
            a

> # Application of CoC1 to our model, the result is the reduced system ReducedODEs
> ApplyChangeOfCoord (ODEs, CoC1);
   d                    2       d                    2
  [-- x(t) = 1 - x(t) + k2 x(t)  y(t), -- y(t) = b - k2 x(t)  y(t)]
   dt                           dt

> # The final change of coordinates giving the relationships
> # between the original system ODEs and the last simplified system FinalODEs
> FinalCoC := ComposeChangeOfCoord (CoC1, CoC2);
                       2
                    k2 a            x k1       y a k2
  FinalCoC := [b = b/a, k2 = -----, t = t k1, x = ----, y = ------]
                       3            a            2
                    k1                         k1

> # Application of FinalCoC to our model, the result is the last simplified system FinalODEs
> ApplyChangeOfCoord (ODEs, FinalCoC);
   d                   2        d                     2
  [-- x(t) = 1 - x(t) + x(t)  y(t), -- y(t) = -k2 (-b + x(t)  y(t))]
   dt                            dt
```

3.4 Qualitative Analysis

Functions: `JacobianMatrix`,
  `HopfBifurcationConditions`, `SteadyPointSystem`

The last part of **MABSys** involves some qualitative analysis tools for computing steady points, Hopf bifurcation conditions, etc. There is a large literature for this kind of studies, among which one can cite [10,9,8] for continuous dynamical systems. The qualitative analysis functions of **MABSys** don't come with new ideas. Some of them already exist in some **Maple** packages or use **Maple** commands in their implementations.

Example 9. Let us analyze the oscillation behavior of Van der Pol oscillator of the form:
$$\begin{cases} \frac{\mathrm{d}x}{\mathrm{d}t} = \left(1 - y^2\right)\mu x - y, \\ \frac{\mathrm{d}y}{\mathrm{d}t} = x \end{cases} \tag{9}$$

via Routh-Hurwitz criterion (see § I.13 of [9]) using **MABSys**. First, we define the differential system and deduce the associated algebraic system that defines its steady points.

```
> # The definition of the system of ODEs.
> ODEs := [diff(x(t),t)=mu*(1-y(t)^2)*x(t)-y(t),diff(y(t),t)=x(t)];
          d                         2                     d
  ODEs := [-- x(t) = mu (1 - y(t) ) x(t) - y(t), -- y(t) = x(t)]
          dt                                     dt

> # Algebraic equations defining its steady points.
> SteadyPoint := SteadyPointSystem(ODEs);
                            2
    SteadyPoint := [mu (1 - y ) x - y, x]

> # The steady point of the system.
> SP := [x=0, y=0];
    SP := [x = 0, y = 0]
```

The behavior of a system of ODEs, at the neighborhood of a given steady point, is classified by looking the signs of the eigenvalues of the Jacobian evaluated at this point (see § 1.3 of [10]). This matrix follows.

```
> # The associated Jacobian matrix.
> J := JacobianMatrix(ODEs, statevars=[x,y]);
         [            2                     ]
    J := [-mu (-1 + y )    -2 mu y x - 1]
         [                                 ]
         [      1                0         ]

> J0 := subs(SP, J);
           [mu    -1]
    J0 := [          ]
           [1      0]
```

The Routh-Hurwitz criterion applied on the characteristic polynomial of the evaluated Jacobian gives necessary conditions so that a Hopf bifurcation may occur. In the neighborhood of a Hopf bifurcation, a stable steady point of the system gives birth to a small stable limit cycle. Here is the **MABSys** commands that compute these conditions:

```
> P := LinearAlgebra:-CharacteristicPolynomial(J0, lambda);
              2
    P := lambda  - mu lambda + 1

> # Necessary conditions so that a Hopf bifurcation can happen
> pos, zero, neg := HopfBifurcationConditions(CP, lambda);
    pos, zero, neg := [1], -mu, -1
```

According to this result, conditions to satisfy follow:
$$1 > 0, \quad -\mu = 0, \quad -1 < 0. \tag{10}$$

Because the first and the third conditions are obvious, one can conclude that a Hopf bifurcation may occur for the Van der Pol oscillator if $\mu = 0$.

4   Tyson's Negative Feedback Oscillator

In [30], authors show how to embed simple signaling pathways in networks using feedbacks to generate more complex behaviors in non-linear control systems. This section is devoted to the analysis of, what we call, Tyson's negative feedback oscillator figure 2.a taken from this paper. Such negative feedback loops are widely used in modeling in biology domain to cause oscillations (see [7,17,28]). We illustrate how MABSys, especially its exact simplification part, could help to predict numerical values that engender oscillation of this particular system and also to clarify the influence of each parameter on this behavior.

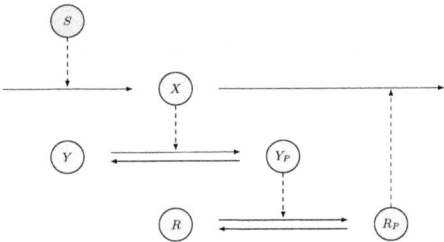

Fig. 3. Tyson's negative feedback oscillator

The figure 3 shows the interactions in the negative-feedback control loop of our study. S is the signal, X, Y, R, Y_P and R_P are other components of the signaling network. The negative feedback loop is closed by R_P activating the degradation of X. The kinetic equations corresponding to this diagram follow:

$$
\begin{cases}
\dfrac{\mathrm{d}X}{\mathrm{d}t} = k_0 + k_1\, S - k_2\, X - k_2'\, R_P\, X, \\[2mm]
\dfrac{\mathrm{d}Y_P}{\mathrm{d}t} = \dfrac{k_3\, X\,(Y_T - Y_P)}{K_{m3} + Y_T - Y_P} - \dfrac{k_4\, Y_P}{K_{m4} + Y_P}, \\[2mm]
\dfrac{\mathrm{d}R_P}{\mathrm{d}t} = \dfrac{k_5\, Y_P\,(R_T - R_P)}{K_{m5} + R_T - R_P} - \dfrac{k_6\, R_P}{K_{m6} + R_P}
\end{cases}
\tag{11}
$$

where $Y_T = Y + Y_P$ and $R_T = R + R_P$ are total concentrations of the associated molecules thus constants. In [30] the studied oscillations arise by the mechanism of Hopf bifurcation. Except for the signal strength S, that varies between two critical values, all parameters are taken fixed. The question arises whether the oscillations occur for other set of parameter values, and if so, how one can find some of them algorithmically. To answer this question we engage in the symbolic simplification of the mathematical model (11) and then we discuss different approaches.

4.1 Symbolic Simplification

The symbolic simplification methods implemented in **MABSys** let us construct an equivalent but simpler model that keeps same qualitative behaviors as the original model. Assuming that the parameters are positive i.e. in $\mathbb{R}_+^\star$, we impose a bijection between the positive solutions of these original and simplified models. Moreover, the exact relationships between the old and the new coordinates permit us to predict sets of numerical values of kinetic constants for which oscillatory behavior is observed.

The model (11) possesses 15 parameters and 3 state variables. For a qualitative analysis, such a system arises complexity problems for symbolic approaches and has too large exploration domains for numerical approaches. That is why, using **MABSys** and its functions of polynomial complexity in the input size, we can find an equivalent system with less and better organized parameters.

Here are the **MABSys** commands and their outputs used to perform the reduction (see § 3.3) and the semi-rectification (see § 3.3) methods. We assume that all parameters are positive and give them by decreasing order of preference for both simplifications. Note that eliminated or factorized parameters can change according their order in function calls. The state variables, given in the second argument, are supposed to not be used in the new parameters expressions.

```
> # Tyson's negative feedback model
> ODEs := [diff(X(t),t)=k0+k1*S-k2*X(t)-k2p*RP(t)*X(t),
>          diff(YP(t),t)=(k3*X(t)*(YT-YP(t)))/(Km3+YT-YP(t))-(k4*YP(t))/(Km4+YP(t)),
>          diff(RP(t),t)=(k5*YP(t)*(RT-RP(t)))/(Km5+RT-RP(t))-(k6*RP(t))/(Km6+RP(t))]:

> # Reduction
> out := InvariantizeByScalings(ODEs,[k1,k3,k5,Km3,Km5,k4,k6,k2p,Km4,Km6,k2,YT,RT,S],[X,YP,RP,k0]):
> InterODEs := out[1]:
> CoC1 := out[2]:
> FreedParams1 := out[3];
                              FreedParams1 := [k1, k3, k5, Km3, Km5]
> # Semi-rectification
> out := SemiRectifySteadyPoints(InterODEs,[k2p,k2,Km4,Km6,k4,k6,YT,RT,S],[X,YP,RP,k0]):
> FinalODEs := out[1,1]:
> CoC2 := out[1,3]:
> FreedParams2 := out[1,4];
                              FreedParams2 := [k2p, k4]
> # Results
> FinalChangeOfCoord := ComposeChangeOfCoord(CoC1,CoC2);
                                               2
                      k4 Km5          k6          k2p Km5         Km4        Km6        k0 k3
FinalChangeOfCoord := [k4 = -------, k6 = ------, k2p = --------, Km4 = ---, Km6 = ---, k0 = ----------,
                        2            Km3 k5         Km3 k5          Km3        Km5       Km5 k2p k4
                      Km3 k5

         k2           YT        RT        S k1 k3        t Km3 k5      X k3       YP       RP
    k2 = -------, YT = ---, RT = ---, S = ----------, t = --------, X = ----, YP = ---, RP = ---]
         Km5 k2p       Km3       Km5      Km5 k2p k4       Km5          k4        Km3      Km5

> FinalODEs;
  d
  [-- X(t) = (k0 + S - k2 X(t) - RP(t) X(t)) k2p,
  dt

  d           k4 (-X(t) YT Km4 - X(t) YT YP(t) + X(t) YP(t) Km4 + X(t) YP(t)  + YP(t) + YT YP(t) - YP(t) )
  -- YP(t) = ---------------------------------------------------------------------------------------------,
  dt                                              (-1 - YT + YP(t)) (Km4 + YP(t))

  d           -YP(t) RT Km6- YP(t) RT RP(t)+ YP(t) RP(t) Km6+ YP(t) RP(t)  + k6 RP(t)+ k6 RP(t) RT- k6 RP(t)
  -- RP(t) = ---------------------------------------------------------------------------------------------]
  dt                                              (-1 - RT + RP(t)) (Km6 + RP(t))
```

It took less then 2 seconds to find an equivalent system where the number of parameters of the whole differential system is decreased by 5 and the number of parameters defining the location of steady points is decreased by 7. According to the outputs of the `MABSys` commands, this equivalent system written in the new coordinates set $\widetilde{Z}$ follows in the factorized form:

$$\begin{cases} \dfrac{\mathrm{d}\widetilde{X}}{\mathrm{d}\widetilde{t}} = \left(\widetilde{k}_0 + \widetilde{S} - \widetilde{k}_2\,\widetilde{X} - \widetilde{R}_P\,\widetilde{X} \right) \widetilde{k}_2', \\[2ex] \dfrac{\mathrm{d}\widetilde{Y}_P}{\mathrm{d}\widetilde{t}} = \left(\dfrac{\widetilde{X}\left(\widetilde{Y}_T - \widetilde{Y}_P\right)}{1 + \widetilde{Y}_T - \widetilde{Y}_P} - \dfrac{\widetilde{Y}_P}{\widetilde{K}_{m4} + \widetilde{Y}_P} \right) \widetilde{k}_4, \\[2ex] \dfrac{\mathrm{d}\widetilde{R}_P}{\mathrm{d}\widetilde{t}} = \dfrac{\widetilde{Y}_P\left(\widetilde{R}_T - \widetilde{R}_P\right)}{1 + \widetilde{R}_T - \widetilde{R}_P} - \dfrac{\widetilde{k}_6\,\widetilde{R}_P}{\widetilde{K}_{m6} + \widetilde{R}_P}. \end{cases} \qquad (12)$$

Again, according to the above outputs, the new coordinates are described w.r.t. the old ones as follows:

$$\begin{aligned} &\widetilde{k}_4 = \tfrac{k_4\,K_{m5}}{K_{m3}^2\,k_5}, \quad \widetilde{k}_6 = \tfrac{k_6}{K_{m3}\,k_5}, \quad \widetilde{k}_2' = \tfrac{k_2'\,K_{m5}^2}{K_{m3}\,k_5}, \quad \widetilde{K}_{m4} = \tfrac{K_{m4}}{K_{m3}}, \quad \widetilde{K}_{m6} = \tfrac{K_{m6}}{K_{m5}}, \\ &\widetilde{k}_0 = \tfrac{k_0\,k_3}{K_{m5}\,k_2'\,k_4}, \quad \widetilde{k}_2 = \tfrac{k_2}{K_{m5}\,k_2'}, \quad \widetilde{Y}_T = \tfrac{Y_T}{K_{m3}}, \quad \widetilde{R}_T = \tfrac{R_T}{K_{m5}}, \quad \widetilde{S} = \tfrac{S\,k_1\,k_3}{K_{m5}\,k_2'\,k_4}, \\ &\widetilde{t} = \tfrac{t\,K_{m3}\,k_5}{K_{m5}}, \quad \widetilde{X} = \tfrac{X\,k_3}{k_4}, \quad \widetilde{Y}_P = \tfrac{Y_P}{K_{m3}}, \quad \widetilde{R}_P = \tfrac{R_P}{K_{m5}}, \\ &\widetilde{k}_1 = k_1, \quad\qquad \widetilde{k}_3 = k_3, \quad\quad \widetilde{K}_{m3} = K_{m3}, \widetilde{k}_5 = k_5, \quad \widetilde{K}_{m5} = K_{m5}. \end{aligned} \qquad (13)$$

Remark that the parameters $\widetilde{k}_1, \widetilde{k}_3, \widetilde{k}_5, \widetilde{K}_{m3}, \widetilde{K}_{m5}$ do not appear in the equations of the system (12) and the parameters $\widetilde{k}_1, \widetilde{k}_3, \widetilde{k}_5, \widetilde{K}_{m3}, \widetilde{K}_{m5}, \widetilde{k}_2', \widetilde{k}_4$ do not influence the location of its steady points. These properties allow to considerably facilitate further analyses and to deduce supplementary information about the original system dynamics.

4.2   Discussion

The previous exact simplifications help the analysis of the studied system by constructing an equivalent system with less parameters and same qualitative behaviors under some positivity assumptions. This fact eases considerably to go further on the comprehension of the biological phenomena. Computed change of coordinates between the original and the simplified systems are effective to find useful numerical values. In the sequel we discuss two such cases.

Oscillation values. The analysis of the simplified model (12) yields information about the original model (11) via the change of coordinates (13). The parameter values which let the simplified model oscillate can be carried back to the original model. Since the number of parameters of the new system is less then the original one, we get a *set* of parameters values that can be adjusted by free numerical values. In our case, there are 5 free values, as much as the number of eliminated parameters. The following set:

$$\begin{aligned} &\widetilde{k}_4 = 200,\ \widetilde{k}_6 = 50,\ \widetilde{k}_2' = 1, \quad \widetilde{K}_{m4} = 1,\ \widetilde{K}_{m6} = 1, \\ &\widetilde{k}_0 = 0,\quad \widetilde{k}_2 = 0.1,\ \widetilde{Y}_T = 100,\ \widetilde{R}_T = 100,\ \widetilde{S} = 10 \end{aligned} \qquad (14)$$

with suitable initial conditions for the state variables causes the oscillation of the simplified model (12). By the help of the new coordinates expressions, we can conclude that any parameter values set satisfying following relations allow the oscillation of the original system:

$$200 = \frac{k_4 K_{m5}}{K_{m3}^2 k_5}, 50 = \frac{k_6}{K_{m3} k_5}, \ 1 = \frac{k_2' K_{m5}^2}{K_{m3} k_5}, 1 = \frac{K_{m4}}{K_{m3}}, \quad 1 = \frac{K_{m6}}{K_{m5}},$$
$$0 = \frac{k_0 k_3}{K_{m5} k_2' k_4}, \ 0.1 = \frac{k_2}{K_{m5} k_2'}, \ 100 = \frac{Y_T}{K_{m3}}, \ 100 = \frac{R_T}{K_{m5}}, \ 10 = \frac{S k_1 k_3}{K_{m5} k_2' k_4}. \tag{15}$$

Five of the parameters, not necessarily the eliminated ones, can take any desired value satisfying above equalities. The change of the timescale and the initial conditions is only important for the coherence of simulations. This phenomenon helps to adapt the system information in order to reproduce the oscillation behavior.

For example, let us consider the specializations $k_1 = 1$, $k_2' = 10$, $k_4 = 0.2$, $Y_T = 1$ and $K_{m6} = 0.01$. The above relations impose (except k_3 because of the null value of $\tilde{k}_0$):

$$\begin{aligned} k_4 &= 0.2, \ k_6 = 0.05, \ k_2' = 10, & K_{m4} &= 0.01, \ K_{m6} = 0.01, \\ k_0 &= 0, \quad k_2 = 0.01, \ Y_T = 1, & R_T &= 1, \qquad S = 2, \\ k_1 &= 1, \quad k_3 = 0.1, \ K_{m3} = 0.01, \ k_5 = 0.1, & K_{m5} &= 0.01. \end{aligned} \tag{16}$$

We recognize the parameter values given in [30]. If we perform the specialization $S = 8$, $k_6 = 1$, $R_T = 10$, $K_{m5} = 0.1$ and $k_4 = 4$ then the above relations impose:

$$\begin{aligned} k_4 &= 4, \ k_6 = 1, & k_2' &= 0.2, & K_{m4} &= 0.1, \ K_{m6} = 0.1, \\ k_0 &= 0, \ k_2 = 0.002, \ Y_T = 10, & R_T &= 10, & S &= 8, \\ k_1 &= 1, \ k_3 = 0.1, & K_{m3} &= 0.1, \ k_5 = 0.2, & K_{m5} &= 0.1. \end{aligned} \tag{17}$$

These values oscillates also the original model of Tyson's negative feedback loop. Remark that the same logic works for any other parameter values than (14) that cause the oscillation of the simplified model (12).

Influence of parameters. One of the contributions of these simplification procedures is to show which quantities really influence the dynamics of the studied system. For example, in (12) we see clearly that the steady points of the system depend on 8 parameters $\tilde{k}_0, \tilde{S}, \tilde{k}_2, \tilde{Y}_T, \tilde{K}_{m4}, \tilde{R}_T, \tilde{k}_6, \tilde{K}_{m6}$. If we return back to the original coordinates, in fact we deduce that what really matters is the following equalities:

$$\begin{aligned} \tilde{k}_0 &= \frac{k_0 k_3}{K_{m5} k_2' k_4}, \ \tilde{S} = \frac{S k_1 k_3}{K_{m5} k_2' k_4}, \ \tilde{k}_2 = \frac{k_2}{K_{m5} k_2'}, \ \tilde{Y}_T = \frac{Y_T}{K_{m3}}, \\ \tilde{K}_{m4} &= \frac{K_{m4}}{K_{m3}}, \quad \tilde{R}_T = \frac{R_T}{K_{m5}}, \quad \tilde{k}_6 = \frac{k_6}{K_{m3} k_5}, \ \tilde{K}_{m6} = \frac{K_{m6}}{K_{m5}} \end{aligned} \tag{18}$$

that come from the change of coordinates computed by MABSys. Certainly, it is still possible to ripen these dependency relations but for now, our methods give a non-negligible start.

Let us illustrate how one can use this information concretely for the numerical values. For instance, with no preceding knowledge, looking for numerical values that change the location of steady points of (11) require a "blind" exploration of the domain in which the parameters live. However we can improve this research using (18). Suppose that we have the location of a steady point $\left(\widetilde{X}_0, \widetilde{Y}_{P0}, \widetilde{R}_{P0} \right)$ of (12) for the following numerical values:

$$
\begin{aligned}
&\widetilde{k}_0 = 0, \quad \widetilde{S} = 5, \quad \widetilde{k}_2 = 0.1, \, \widetilde{Y}_T = 50, \\
&\widetilde{K}_{m4} = 2, \, \widetilde{R}_T = 50, \, \widetilde{k}_6 = 20, \, \widetilde{K}_{m6} = 1.
\end{aligned}
\tag{19}
$$

Any original parameters values satisfying these equalities, for example, each of the following sets:

$$
\begin{aligned}
&k_4 = 0.125, \, k_6 = 4, \quad k'_2 = 2, \quad\;\; K_{m4} = 0.4, \, K_{m6} = 0.2, \\
&k_0 = 0, \qquad k_2 = 0.04, \, Y_T = 10, \quad R_T = 10, \quad S = 2.5, \\
&k_1 = 1, \qquad k_3 = 0.1, \quad K_{m3} = 0.2, \, k_5 = 1, \qquad K_{m5} = 0.2
\end{aligned}
\tag{20a}
$$

$$
\begin{aligned}
&k_4 = 4, \, k_6 = 2, \quad k'_2 = 1, \quad\; K_{m4} = 0.2, \, K_{m6} = 0.1, \\
&k_0 = 0, \, k_2 = 0.01, \, Y_T = 5, \quad R_T = 5, \quad S = 1, \\
&k_1 = 5, \, k_3 = 0.4, \quad K_{m3} = 0.1, \, k_5 = 1, \qquad K_{m5} = 0.1
\end{aligned}
\tag{20b}
$$

$$
\begin{aligned}
&k_4 = 2, \, k_6 = 0.4, \, k'_2 = 10, \quad\; K_{m4} = 0.02, \, K_{m6} = 0.01, \\
&k_0 = 0, \, k_2 = 0.01, \, Y_T = 0.5, \quad R_T = 0.5, \quad S = 2, \\
&k_1 = 5, \, k_3 = 0.1, \quad K_{m3} = 0.01, \, k_5 = 2, \qquad K_{m5} = 0.01
\end{aligned}
\tag{20c}
$$

yields predictable locations of steady points for the original system (11). These locations (X_0, Y_{P0}, R_{P0}) can be deduced directly by a simple transformation coming from (13):

$$
X_0 = \frac{k_4 \, \widetilde{X}_0}{k_3}, \quad Y_{P0} = K_{m3} \, \widetilde{Y}_{P0}, \quad R_{P0} = K_{m5} \, \widetilde{R}_{P0}.
\tag{21}
$$

5   Installation of MABSys

The software is available for download with the associated help pages [20]. It requires at least Maple 11. Because MABSys uses some types and functions of the package ExpandedLiePointSymmetry, you need to install also this last one. First, download the libraries of ExpandedLiePointSymmetry in a directory named, for example, ~/softwares/ExpandedLiePointSymmetry/ and the libraries of MABSys in a directory named ~/softwares/MABSys/. Second, make them visible to Maple. The global variable libname must be updated. You can do it once forever in your .mapleinit file or every time you open a Maple session as:

```
> libname := libname, "~/softwares/ExpandedLiePointSymmetry", "~/softwares/MABSys/":
```

Once the libname has been properly set, the package can be loaded and you can enjoy it by the following command.

```
> with(MABSys);
[ApplyChangeOfCoord, ComposeChangeOfCoord, ConservationLaws, Equilibria, EvaluateSolution,
 GetAllRateConstants, GetAllSubstratesName, GetFastReactions, GetReactionProducts, GetReactionRate,
 GetReactionReactants, GetSlowReactions, HopfBifurcationConditions, HurwitzDeterminants, HurwitzMatrix,
 InvariantizeByScalings, InverseChangeOfCoord, IsReactionFast, JacobianMatrix, ModelReduce, NewReaction,
 NumericalSolution, PlotSolution, RateVector, ReactionSystem2ODEs, SemiRectifyAlgebraicSystem,
 SemiRectifySemiAlgebraicSystem, SemiRectifySteadyPoints, SteadyPointSystem, StoichiometricMatrix]
```

The online help is also loaded and is accessible by ?MABSys.

6 Conclusion and Perspectives

We have presented our MABSys package and illustrated its main functions. Even if MABSys' pilot implementation is in Maple, one of the perspectives is to have a computer algebra software independent package. Other open softwares or programming languages such as C can be preferred in the future to reach more people. Moreover, we seek to improve our modeling and simplification procedures along with our qualitative analysis tools.

References

1. Boulier, F., Lefranc, M., Lemaire, F., Morant, P.-E.: Applying a Rigorous Quasi-Steady State Approximation Method for Proving the Absence of Oscillations in Models of Genetic Circuits. In: Horimoto, K., Regensburger, G., Rosenkranz, M., Yoshida, H. (eds.) AB 2008. LNCS, vol. 5147, pp. 56–64. Springer, Heidelberg (2008), http://hal.archives-ouvertes.fr/hal-00213327/
2. Boulier, F.: Réécriture algébrique dans les systèmes d'équations différentielles polynomiales en vue d'applications dans les Sciences du Vivant. H497. Université de Lille 1, LIFL, 59655 Villeneve d'Ascq France, Mémoire d'Habilitation à Diriger des Recherches (May 2006)
3. Boulier, F.: Differential Elimination and Biological Modelling. In: Rosenkranz, M., Wang, D. (eds.) Gröbner Bases in Symbolic Analysis Workshop D2.2 of the Special Semester on Gröbner Bases and Related Methods, Hagenberg Autriche. Radon Series Comp. Appl. Math, vol. 2, pp. 111–139. De Gruyter (2007)
4. Boulier, F., Lefranc, M., Lemaire, F., Morant, P.-E.: Model Reduction of Chemical Reaction Systems using Elimination. In: MACIS (2007), http://hal.archives-ouvertes.fr/hal-00184558/fr
5. Boulier, F., Lefranc, M., Lemaire, F., Morant, P.-E., Ürgüplü, A.: On Proving the Absence of Oscillations in Models of Genetic Circuits. In: Anai, H., Horimoto, K., Kutsia, T. (eds.) AB 2007. LNCS, vol. 4545, pp. 66–80. Springer, Heidelberg (2007), http://hal.archives-ouvertes.fr/hal-00139667
6. Briggs, G.E., Haldane, J.B.S.: A note on the kinetics of enzyme action. Biochemical Journal 19, 338–339 (1925)
7. Griffith, J.S.: Mathematics of Cellular Control Processes. I. Negative Feedback to One Gene. Journal of Theoretical Biology 20, 202–208 (1968)
8. Guckenheimer, J., Myers, M., Sturmfels, B.: Computing Hopf Bifurcations I. SIAM Journal on Numerical Analysis (1997)
9. Hairer, E., Norsett, S.P., Wanner, G.: Solving ordinary differential equations I: non-stiff problems, 2nd revised edn. Springer-Verlag New York, Inc., New York (1993)
10. Hale, J., Koçak, H.: Dynamics and Bifurcations. Texts in Applied Mathematics, vol. 3. Springer, New York (1991)
11. Heck, A.: Introduction to Maple, 3rd edn. Springer, Heidelberg (2003) ISBN 0-387-00230-8

12. Henri, V.: Lois générales de l'Action des Diastases. Hermann, Paris (1903)
13. Hubert, É.: AIDA Maple package: Algebraic Invariants and their Differential Algebras (2007)
14. Hubert, É., Sedoglavic, A.: Polynomial Time Nondimensionalisation of Ordinary Differential Equations via their Lie Point Symmetries. Internal Report (2006)
15. Hucka, M., Keating, S.M., Shapiro, B.E., Jouraku, A., Tadeo, L.: SBML (The Systems Biology Markup Language) (2003), http://sbml.org
16. Khanin, R.: Dimensional Analysis in Computer Algebra. In: Mourrain, B. (ed.) Proceedings of the 2001 International Symposium on Symbolic and Algebraic Computation, London, Ontario, Canada, July 22-25, pp. 201–208. ACM, ACM press (2001)
17. Kholodenko, B.N.: Negative feedback and ultrasensitivity can bring about oscillations in the mitogen-activated protein kinase cascades. European Journal of Biochemistry 267, 1583–1588 (2000)
18. Lemaire, F., Maza, M.M., Xie, Y.: The RegularChains library in MAPLE 10. In: Kotsireas, I.S. (ed.) The MAPLE Conference, pp. 355–368 (2005)
19. Lemaire, F., Ürgüplü, A.: A Method for Semi-Rectifying Algebraic and Differential Systems using Scaling type Lie Point Symmetries with Linear Algebra. In: Proceedings of ISSAC (2010) (to appear)
20. Lemaire, F., Ürgüplü, A.: Modeling and Analysis of Biological Systems. Maple Package (2008), www.lifl.fr/~urguplu
21. Mansfield, E.: Indiff: a MAPLE package for over determined differential systems with Lie symmetry (2001)
22. Michaëlis, L., Menten, M.: Die Kinetik der Invertinwirkung (the kinetics of invertase activity). Biochemische Zeitschrift 49, 333–369 (1973), Partial english translation, http://web.lemoyne.edu/~giunta/menten.html
23. Murray, J.D.: Mathematical Biology. Interdisciplinary Applied Mathematics, vol. 17. Springer, Heidelberg (2002)
24. Okino, M.S., Mavrovouniotis, M.L.: Simplification of Mathematical Models of Chemical Reaction Systems. Chemical Reviews 98(2), 391–408 (1998)
25. Ritt, J.F.: Differential Algebra. American Mathematical Society Colloquium Publications, vol. XXXIII. AMS, New York (1950), http://www.ams.org/online_bks/coll33
26. Sedoglavic, A.: Reduction of Algebraic Parametric Systems by Rectification of Their Affine Expanded Lie Symmetries. In: Anai, H., Horimoto, K., Kutsia, T. (eds.) AB 2007. LNCS, vol. 4545, pp. 277–291. Springer, Heidelberg (2007), http://hal.inria.fr/inria-00120991
27. Sedoglavic, A., Ürgüplü, A.: Expanded Lie Point Symmetry, Maple package (2007), http://www.lifl.fr/~urguplu
28. Smolen, P., Baxter, D.A., Byrne, J.H.: Modeling circadian oscillations with interlocking positive and negative feedback loops. Journal of Neuroscience 21, 6644–6656 (2001)
29. Szallasi, Z., Stelling, J., Periwal, V. (eds.): System Modeling in Cellular Biology. The MIT Press, Cambridge (2006)
30. Tyson, J.J., Chen, K.C., Novak, B.: Sniffers, buzzers, toggles and blinkers: dynamics of regulatory and signaling pathways in the cell. Current Opinion in Cell Biology 15, 221–231 (2003)
31. Ürgüplü, A.: Contribution to Symbolic Effective Qualitative Analysis of Dynamical Systems; Application to Biochemical Reaction Networks. PhD thesis, University of Lille 1 (January 13, 2010)
32. Vora, N., Daoutidis, P.: Nonlinear model reduction of chemical reaction systems. AIChE (American Institute of Chemical Engineers) Journal 47(10), 2320–2332 (2001)

Models of Stochastic Gene Expression
and Weyl Algebra

Samuel Vidal[1], Michel Petitot[2], François Boulier[2],
François Lemaire[2], and Céline Kuttler[2,3]

[1] Univ. Lille I
Lab. Paul Painlevé
samuel.vidal@math.univ-lille1.fr
[2] Univ. Lille I
LIFL
{michel.petitot,francois.boulier,
francois.lemaire,celine.kuttler}@lifl.fr
[3] Univ. Lille I
IRI

Abstract. This paper presents a symbolic algorithm for computing the ODE systems which describe the evolution of the moments associated to a chemical reaction system, considered from a stochastic point of view. The algorithm, which is formulated in the Weyl algebra, seems more efficient than the corresponding method, based on partial derivatives. In particular, an efficient method for handling conservation laws is presented. The output of the algorithm can be used for a further investigation of the system behaviour, by numerical methods. Relevant examples are carried out.

Keywords: Stochastic models, Weyl algebra, Generating series.

1   Introduction

This paper is concerned with the modeling of gene regulatory networks by chemical reaction systems, from a stochastic point of view. To such systems, it is possible to associate a time varying random variable which counts the numbers of molecules of the various chemical species. It is well-known that the evolution, over time t, of the moments (mean, variance, covariance) of this random variable, may be described by a system of ordinary differential equations (ODEs), at least for first order chemical reaction systems [6,16]. See also [13] for an introduction to these topics. These systems of ODEs can be built from the probability generating function associated to a given chemical reaction system, by performing, essentially, the three following steps:

1. compute a Schrödinger equation analog [4, Eq. (5.60)] for the probability generating function $\phi(t, z)$, where t denotes the time and z denotes a vector of formal variables ;

K. Horimoto, M. Nakatsu, and N. Popov (Eds.): ANB 2011, LNCS 6479, pp. 76–97, 2012.
© Springer-Verlag Berlin Heidelberg 2012

2. compute iterated derivatives of this equation, with respect to t ;
3. evaluate the differentiated equation at $z = 1$.

This paper shows how to build these ODE systems by using Weyl algebra methods. The idea consists in formulating the Schrödinger equation analog using Euler derivation operators (of the form $z \, \partial/\partial z$) instead of more traditional partial derivatives (of the form $\partial/\partial z$). As far as we know, the use of Weyl algebra in this context is new. It leads to a new algorithm which seems more efficient than the equivalent method, based on the use of partial derivatives. This last claim is not proved in this paper. It was suggested to us by the following observations. The formulation in the Weyl algebra permits to:

1. "combine in one step" steps 2 and 3 above, and thereby reduce the expression swell produced by step 2 (Formula (14) in Proposition 2) ;
2. prove the Formula (14), which we find more compact and simpler than the formulas in [6,16].
3. allows to encode, in the Schrödinger equation analog, the linear conservation laws of the system (Algorithm of Section 5.1), and thereby take advantage of them at the very first step of the method ;

A software prototype has been developed by the second author in the MAPLE computer algebra software. Supplementary related data are available at the url http://www.lifl.fr/~petitot/recherche/exposes/ANB2010.

The paper is organized as follows. In Section 2, the classical theory is recalled. The material can essentially be found (often piecewise) in many texts such as [4, chapter 5] or [8]. We feel the need to recall it in order to avoid confusions, since, depending on slight variants of the underlying assumptions, or slight variants of notations, different formula may be obtained. Our presentation is based on stochastic Petri nets [14,17]. In Section 3, the Weyl algebra and Euler operators are introduced, and the Schrödinger equation analog is reformulated, in this setting. In Section 4, the construction of the ODE system for the moments, from the differential operators, is explained (Proposition 1) and the algorithm is stated. In Section 5, the method for simplifying the Schrödinger equation analog using the conservation laws is provided. Section 6 provides a, new, combined formula for steps 1 and 2 (Formula (14) in Proposition 2). Some properties of the algorithm, which depend on the order of the chemical reaction system under study are explored in Section 7. In particular, some well-known results (related to compartmental models) are recovered from the Weyl algebra theory. Some examples are carried out in Section 8. The problem of the "infinite cascade" is studied for second order systems. Many parts of this article can be found in [15, Annexe A].

2   The Classical Theory

2.1   Chemical Reactions Systems

Definition 1. *A chemical reaction system is given by a set of chemical species* $(R_1, R_2, \ldots, R_n)$, *and, a set of chemical reactions, of the following form, where*

$\alpha = (\alpha_1, \alpha_2, \ldots, \alpha_n)$, $\beta = (\beta_1, \beta_2, \ldots, \beta_n)$ *are multi-indices of nonnegative integers and c is a positive real valued* kinetic constant[1].

$$\alpha_1 R_1 + \cdots + \alpha_n R_n \xrightarrow{c} \beta_1 R_1 + \cdots + \beta_n R_n \qquad (1)$$

Such a system, denoted $\mathcal{R}$, involves several reactions in general ; therefore it is a finite set of triples $(c, \alpha, \beta) \in \mathbb{R}_{>0} \times \mathbb{N}^n \times \mathbb{N}^n$. The state of the system $\nu = (\nu_1, \nu_2, \ldots, \nu_n) \in \mathbb{N}^n$ is the number of molecules of the n chemical species at a given time. When reaction (1) occurs, the state vector instantaneously changes from the current value ν to the new value $\nu' = \nu - \alpha + \beta$. The vector α indicates the quantities of the various species being consumed by the reaction while the vector β indicates the produced quantities. For a reaction to occur it is necessary that $\alpha \leq \nu$, i.e. $\alpha_i \leq \nu_i$ for $i = 1 \ldots n$. An example is provided by the following system, which is encoded by the set $\mathcal{R} = \{(\lambda, (2, 1, 0, 0), (0, 0, 1, 0)), (\mu, (0, 0, 1, 0), (0, 0, 0, 3))\}$.

$$2R_1 + R_2 \xrightarrow{\lambda} R_3 , \qquad R_3 \xrightarrow{\mu} 3R_4 . \qquad (2)$$

2.2 Stochastic Petri Nets

Definition 2. *A Petri net is a finite directed bipartite graph. Its nodes represent transitions (i.e. events that may occur, represented by boxes) and places (i.e. conditions for the events, represented by circles).*

Denote $\mathcal{P} = \{p_1, p_2, \ldots, p_n\}$ the set of places and $\mathcal{T} = \{t_1, t_2, \ldots, t_m\}$ the set of transitions. At every instant, the place p_k, $(1 \leq k \leq n)$ is supposed to contain ν_k indistinguishable tokens. The state vector $\nu = (\nu_1, \nu_2, \ldots, \nu_n) \in \mathbb{N}^n$ is the number of tokens present in the places. One denotes C^- and C^+ the integer matrices where $C^-_{i,j}$ is the number of arrows going from place p_j to transition t_i, and $C^+_{i,j}$ is the number of arrows going from transition t_i to place p_j. The *incidence matrix* is $C = C^+ - C^-$.

Definition 3. *A* stochastic Petri net *is a Petri net endowed with a function* $\rho : \mathcal{T} \longrightarrow \mathbb{R}_{>0}$.

The category of the stochastic Petri nets is strictly equivalent to the category of chemical reaction systems. To each transition $t_i \in \mathcal{T}$, a chemical reaction like (1) is associated. It is coded by a triple $(c, \alpha, \beta) \in \mathbb{R}_{>0} \times \mathbb{N}^n \times \mathbb{N}^n$ with $\alpha = (C^-_{ij})_{j=1\ldots n}$, $\beta = (C^+_{ij})_{j=1\ldots n}$ and $c = \rho(t_i)$. The incidence matrix C is then the transpose of the stoichiometry matrix of the chemical reaction system. The correspondence holds:

$$\text{token} \longleftrightarrow \text{chemical molecule}$$
$$\text{place} \longleftrightarrow \text{chemical species}$$
$$\text{transition} \longleftrightarrow \text{chemical reaction}$$

---

[1] Called *stochastic reaction constant* in [7].

The chemical reaction system (2) corresponds to the Petri net of Figure 1 with the associated matrices

$$C^- = \begin{pmatrix} 2 & 1 & 0 & 0 \\ 0 & 0 & 1 & 0 \end{pmatrix}, \qquad C^+ = \begin{pmatrix} 0 & 0 & 1 & 0 \\ 0 & 0 & 0 & 3 \end{pmatrix}, \qquad C = \begin{pmatrix} -2 & -1 & 1 & 0 \\ 0 & 0 & -1 & 3 \end{pmatrix}.$$

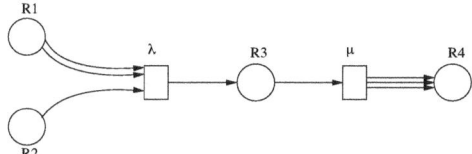

Fig. 1. Petri net of Example (2)

2.3   Markov Chain of the Temporisation and Master Equation

One gives the "standard temporisation" of a stochastic Petri net by associating it with a continuous time Markov chain $\{N(t); \ t \in \mathbb{R}_{\geq 0}\}$ where the vector valued random variable $N(t)$ counts, at time t, the number of tokens in the places of the network. Let $\pi_\nu(t)$ be the probability that the process is in state $N(t) = \nu$ at time t. The *master-equation* [5] of a Markov chain is a linear differential system governing the evolution, over time, of the row vector $\pi(t) = (\pi_\nu(t); \nu \in \mathbb{N}^n)$. It is written as follows:

$$\frac{d}{dt}\pi(t) = \pi(t)\,A \ . \tag{3}$$

For each couple of multi-indices $(\alpha, \nu) \in \mathbb{N}^n \times \mathbb{N}^n$, the following product of binomial coefficients is defined

$$\binom{\nu}{\alpha} = \binom{\nu_1}{\alpha_1}\binom{\nu_2}{\alpha_2} \cdots \binom{\nu_n}{\alpha_n} \ . \tag{4}$$

The Markov chain $\{N(t); \ t \in \mathbb{R}_{\geq 0}\}$ associated to the stochastic Petri net $\mathcal{R}$, is defined on the discrete state space $\mathbb{N}^n$. For each triple $(c, \alpha, \beta) \in \mathcal{R}$ and for each state $\nu \geq \alpha$ an arrow going from ν to $\nu' = \nu - \alpha + \beta$ is built. It is labeled by the transition rate $c\binom{\nu}{\alpha}$. Schematically, this can be written:

$$(\nu + \alpha - \beta) \xrightarrow{c\binom{\nu+\alpha-\beta}{\alpha}} (\nu) \xrightarrow{c\binom{\nu}{\alpha}} (\nu - \alpha + \beta).$$

Assume that the system is in state $\nu \in \mathbb{N}^n$ at time t. The probability that the chemical reaction (c, α, β) occurs within the time range $[t, t + \varepsilon[$ is $\varepsilon\, c\binom{\nu}{\alpha} + o(\varepsilon)$. Then the master-equation (3) takes the form

$$\frac{d}{dt}\pi_\nu(t) = \sum_{(c,\alpha,\beta) \in \mathcal{R}} c\binom{\nu + \alpha - \beta}{\alpha} \pi_{\nu+\alpha-\beta}(t) - c\binom{\nu}{\alpha}\pi_\nu(t). \tag{5}$$

According to Definition (4), the first term of the sum is zero whenever $\beta > \nu$ and the second is zero whenever $\alpha > \nu$. This differential system involves an infinite number of unknowns $\pi_\nu(t)$ constrained by an infinite linear differential system (see Example (8)).

2.4   The Schrödinger Equation Analog

For investigating models analytically we introduce the probability generating function [4, sect. 5.3]

$$\phi(t, z) = \sum_{\nu \geq 0} \pi_\nu(t)\, z^\nu . \tag{6}$$

Note that $\phi(t, z)$ is also equal to $\mathrm{E}\, z^{N(t)}$ (i.e. the mean value of $z^{N(t)}$) since $\mathrm{E}\, z^{N(t)} = \sum_{\nu \geq 0} \mathrm{prob}(z^\nu = z^{N(t)}) z^\nu$. Given any chemical reaction system, it is possible to compute a general equation for ϕ [4, Eq. (5.60)]. This general equation is a Schrödinger equation analog. The differential operator H is a Hamiltonian.

$$\frac{\partial}{\partial t}\phi(z, t) = H\,\phi(z, t) \quad \text{where} \quad H = \sum_{(c,\alpha,\beta) \in \mathcal{R}} \frac{c}{\alpha!}\left(z^\beta - z^\alpha\right)\left(\frac{\partial}{\partial z}\right)^\alpha . \tag{7}$$

The Hamiltonian of Example (2) is

$$H = \frac{1}{2}\lambda\left(z_3 - z_1^2 z_2\right)\frac{\partial^2}{\partial z_1^2}\frac{\partial}{\partial z_2} + \mu\left(z_4^3 - z_3\right)\frac{\partial}{\partial z_3} .$$

By differentiating the probability generating function (6) and evaluating it at $z = 1$, i.e, at $z_1 = \cdots = z_n = 1$, formulas which bind ϕ, the means and the variances of the number of molecules of the chemical species, are obtained. Here are a few examples. A mere evaluation yields: $\phi(t, z)\,|_{z=1} = 1$. Differentiating (6) with respect to any z_i and evaluating at $z = 1$ provides the expected value of the number of molecules of species R_i, i.e. $\mathrm{E}\, N_i(t)$

$$\left(\frac{\partial}{\partial z_i}\phi(z, t)\right)_{|z=1} = \sum_\nu \nu_i\, \pi_\nu(t) = \mathrm{E}\, N_i(t) .$$

Differentiating (6) twice with respect to some fixed z_i and evaluating at $z = 1$ provides a formula featuring the expected value of the square of the number of molecules of R_i, denoted $\mathrm{E}\, N_i(t)^2$, together with $\mathrm{E}\, N_i(t)$:

$$\left(\frac{\partial}{\partial z_i}\frac{\partial}{\partial z_i}\phi(z, t)\right)_{|z=1} = \sum_\nu \nu_i\left(\nu_i - 1\right)\pi_\nu(t)$$

$$= \sum_\nu \nu_i^2\, \pi_\nu(t) - \nu_i\, \pi_\nu(t) = \mathrm{E}\, N_i(t)^2 - \mathrm{E}\, N_i(t) .$$

The variance of the number of molecules of R_i satisfies the well-known formula: $\mathrm{Var}\, N_i(t) = \mathrm{E}\, N_i(t)^2 - (\mathrm{E}\, N_i(t))^2$. The above formula can then be restated using $\mathrm{E}\, N_i(t)$ and $\mathrm{Var}\, N_i(t)$ only. Then, from the arguments above and the Schrödinger equation analog (7), an ODE system for the means and the variances $\mathrm{E}\,N_i(t)$ and $\mathrm{Var}\,N_i(t)$ can be computed. The method is illustrated over the following example:

$$\emptyset \xrightarrow{\lambda} R \qquad R \xrightarrow{\mu} \emptyset . \tag{8}$$

It corresponds to the creation and degradation of mRNA by a unregulated gene. It also corresponds to the $M/M/\infty$ client-server system [5]. The following set of triples $\mathcal{R} = \{ (\lambda, (0), (1)), (\mu, (1), (0)) \}$ provides a description of that system. The Markov chain is described by the transition rates $\lambda\binom{\nu}{0} = \lambda$ and $\mu\binom{\nu}{1} = \mu\nu$ for all $\nu \in \mathbb{N}$. Using the convention $\pi_{-1}(t) = 0$, the master-equation (5) of that

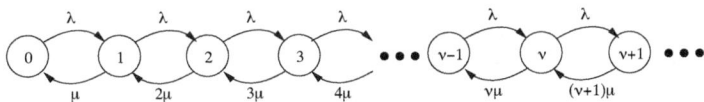

Fig. 2. The Markov chain associated to System (8)

Markov chain is:

$$\frac{d}{dt}\pi_\nu(t) = \lambda\big[\pi_{\nu-1}(t) - \pi_\nu(t)\big] + \mu\big[(\nu+1)\pi_{\nu+1}(t) - \nu\pi_\nu(t)\big] \qquad (\nu \geq 0). \tag{9}$$

The Schrödinger equation analog (7) is:

$$\frac{\partial}{\partial t}\phi(t,z) = \left[\lambda(z-1) + \mu(1-z)\frac{\partial}{\partial z}\right]\phi(t,z) = \lambda\,(z-1)\,\phi(t,z) + \mu\,(1-z)\,\phi_z(t,z) .$$

In order to compute an ODE for the mean $\mathrm{E}\,N(t)$, that relation is differentiated with respect to z,

$$\frac{\partial}{\partial z}\frac{\partial}{\partial t}\phi(t,z) = \lambda\,\big[\phi(t,z) + (z-1)\,\phi_z(t,z)\big] + \mu\,\big[-\phi_z(t,z) + (1-z)\,\phi_{zz}(t,z)\big] .$$

The partial derivatives $\frac{\partial}{\partial t}$ and $\frac{\partial}{\partial z}$ commute. At the point $z = 1$, this equation becomes:

$$\frac{\partial}{\partial t}\phi_z(t,z)_{|z=1} = \lambda - \mu\,\phi_z(t,z)_{|z=1} \qquad \text{i.e.} \qquad \frac{d}{dt}\mathrm{E}\,N(t) = \lambda - \mu\,\mathrm{E}\,N(t) .$$

An ODE for $\mathrm{E}\,N(t)$ is obtained. Similar computations provide an ODE for $\mathrm{Var}\,N(t)$. The initial values are easily obtained since, at $t = 0$, the expected value of $N(t)$ is equal to the initial quantity n_0 of the chemical species and the variance of $N(t)$ is zero. The analysis of that dynamics can be done numerically or symbolically, depending on the instance of the problem.

Using this method, an ODE for any moment $\mathrm{E}\,N^q(t)$, where $q \in \mathbb{N}$, can be computed. In the particular case of *order* 1 chemical reaction systems (Definition 6), the ODE system is finite and exact values of the means and the variances [4, sect. 5.3.3] can be computed. The above example has order 1.

In the general case, the ODE system is infinite, since for any $q \in \mathbb{N}$, the evolutions of the moments of order q depend on moments of higher order (problem of the infinite cascade [6,16]). It is sometimes possible to compute exact values for the means and the variances by *ad hoc* arguments. Otherwise, the ODE system needs to be truncated and provides a more or less usable approximation. The truncation, which is obtained by assuming that random variables $N_i(t)$ are independent, provides the classical deterministic model, which is used when the number of molecules is large.

3   Reformulation in the Weyl Algebra

In this section, the properties of the Weyl algebra [3] needed to understand the algorithm of Section 4 are introduced. Then, the construction of the Hamiltonian (7) is reformulated.

The Weyl algebra $\mathrm{Weyl}_{\mathbb{R}}(z_1, \ldots, z_n) = \mathbb{R}[z_1, z_2, \ldots, z_n][\partial_{z_1}, \partial_{z_2}, \ldots, \partial_{z_n}]$ is the algebra of (polynomial) differential operators defined on the affine algebraic manifold $\mathbb{R}^n$. It is a non-commutative and associative algebra generated by the symbols z_k and ∂_{z_k} for $k = 1 \ldots n$ constrained by the commutation relations $[\partial_{z_i}, z_j] = \partial_{z_i} z_j - z_j \partial_{z_i} = 1$ if $i = j$, and 0 otherwise. The commutator (also called Lie Bracket) between A and B, denoted $[A, B]$, is defined by the relation $[A, B] = AB - BA$. For all $\alpha, \beta \in \mathbb{N}^n$, denote $z^\alpha = z_1^{\alpha_1} z_2^{\alpha_2} \ldots z_n^{\alpha_n}$ and $\partial_z^\beta = \partial_{z_1}^{\beta_1} \partial_{z_2}^{\beta_2} \ldots \partial_{z_n}^{\beta_n}$. For each element $z^\alpha \partial_z^\beta$, define the *degree* $\deg z^\alpha \partial_z^\beta$ as $|\alpha| - |\beta|$, and the *order* $\mathrm{ord}\, z^\alpha \partial_z^\beta$ as $|\beta|$, where $|\alpha| = \alpha_1 + \alpha_2 + \cdots + \alpha_n$ and $|\beta| = \beta_1 + \beta_2 + \cdots + \beta_n$.

We now define the evaluation of a differential operator. Any differential operator $D \in \mathrm{Weyl}_{\mathbb{R}}(z_1, \ldots, z_n)$ can be written in a unique way as a finite sum of terms $f_\nu \partial_z^\nu$ where $\nu \in \mathbb{N}^n$, and where $f_\nu \in \mathbb{R}[z_1, z_2, \ldots, z_n]$ defines a function $\mathbb{R}^n \to \mathbb{R}$. This allows to define the *evaluation* of D at a point $p \in \mathbb{R}^n$:

$$D_{|_p} = \sum_\nu f_\nu(p)\, \partial_z^\nu . \tag{10}$$

The operator $D_{|_p} : \mathbb{R}[z_1, z_2, \ldots, z_n] \to \mathbb{R}$ is defined in a coordinate free way by setting $D_{|_p}(h) = D(h)_{|_p}$ for any function $h \in \mathbb{R}[z_1, z_2, \ldots, z_n]$. The term $D(h)_{|_p}$ denotes the real number obtained by evaluating the function $D(h)$ at p. The evaluation does not commute with the multiplication of the algebra $\mathrm{Weyl}_{\mathbb{R}}(z)$.

We now reformulate the Schrödinger equation analog, using "falling powers" (also called falling factorials) and Euler operators. Falling powers are defined by $x^{\underline{p}} = x(x-1)(x-2) \cdots (x-p+1)$, where $p \in \mathbb{N}$. Then, binomial coefficients can be reformulated as follows: $\binom{x}{p} = (1/p!)\, x^{\underline{p}}$.

Definition 4. *Euler operators are defined by* $\theta_k = z_k\, \partial/\partial z_k$.

For any $\alpha \in \mathbb{N}^n$, one defines $\theta^{\underline{\alpha}} = \theta_1^{\underline{\alpha_1}} \theta_2^{\underline{\alpha_2}} \cdots \theta_n^{\underline{\alpha_n}}$. The next lemma is classical:

Lemma 1. *For any* $\alpha \in \mathbb{N}^n$, *one has* $\theta^{\underline{\alpha}} = z^\alpha \left(\dfrac{\partial}{\partial z} \right)^\alpha$.

Given any chemical reaction system, the Schrödinger equation analog (7) which governs the probability generating function (6) can be formulated in the Weyl algebra as follows. The formula can be justified using Formula (7) and Lemma 1. The Hamiltonian H belongs to $\text{Weyl}_{\mathbb{R}}(z)$. It is a linear operator acting on the formal series in the variables $(z_1, z_2, \ldots, z_n)$:

$$H = \sum_{(c,\alpha,\beta)\in\mathcal{R}} \frac{c}{\alpha!} \left(z^{\beta-\alpha} - 1\right) \theta^{\alpha} . \tag{11}$$

The Hamiltonian of Example (2) can now be written as:

$$H = \frac{1}{2}\lambda \left(\frac{z_3}{z_1^2 z_2} - 1\right) \theta_1(\theta_1 - 1)\theta_2 + \mu \left(\frac{z_4^3}{z_3} - 1\right) \theta_3 .$$

4   The Algorithm

This section presents our algorithm which computes the differential equations satisfied by the moments a stochastic Petri net (up to a certain order). First Proposition 1 is proved. Then our algorithm is stated and proved using Proposition 1.

Since $N(t)$ is a (time dependent) random variable taking values in $\mathbb{N}^n$, it is possible to consider any random variable of the form $f(N(t))$ where the function $f : \mathbb{N}^n \to \mathbb{R}$ is a polynomial function represented by an element $f \in \mathbb{R}[\theta_1, \theta_2, \ldots, \theta_n]$, i.e, a polynomial in the Euler operators. Observe that, by definition of θ_k, the polynomial f is also an element of the algebra $\text{Weyl}_{\mathbb{R}}(z)$. For example, to the polynomial $f = \theta_1^2 \theta_2 + 3\theta_1 \in \mathbb{R}[\theta_1, \theta_2]$, the following objects can be associated: the operator $f(\theta) = \theta_1^2 \theta_2 + 3\theta_1 \in \text{Weyl}_{\mathbb{R}}(z_1, z_2)$; the number $f(\nu) = \nu_1^2 \nu_2 + 3\nu_1 \in \mathbb{R}$; the random variable $f(N(t)) = N_1^2(t)N_2(t) + 3N_1(t)$ defined on $\mathbb{N}^2$ with value in $\mathbb{R}$; the commutator $[f(\theta), H] = f(\theta)H - Hf(\theta)$; and the evaluated commutator, $[f(\theta), H]_{|_{z_1=z_2=1}}$, which is a polynomial in $\mathbb{R}[\theta_1, \theta_2]$. The mean value of the random variable $f(N(t))$ is, by definition, equal to $\mathrm{E}\, f(N(t)) = \sum_{\nu} f(\nu) \pi_{\nu}(t)$.

Lemma 2. *For any polynomial $f \in \mathbb{R}[\theta_1, \theta_2, \ldots, \theta_n]$, interpreted as a differential operator acting on the generating series $\phi(t, z)$, one has the relation:* $\mathrm{E}\, f(N(t)) = f(\theta) \,\mathrm{E}\, z^{N(t)}_{|z=1}.$

Proof. For any element $\nu \in \mathbb{N}^n$, one has $f(\theta)\, z^{\nu} = f(\nu)\, z^{\nu}$. It follows that $f(\theta)\, \phi(t, z) = \sum_{\nu} f(\nu) \pi_{\nu}(t)\, z^{\nu}$. Thus, evaluating at the point $z = 1$ leads to: $f(\theta)\, \mathrm{E}\, z^{N(t)}_{|z=1} = \sum_{\nu} f(\nu) \pi_{\nu}(t) = \mathrm{E}\, f(N(t)).$

Proposition 1. *Given $f \in \mathbb{R}[\theta_1, \theta_2, \ldots, \theta_n]$, denote $f_H(\theta) = [f(\theta), H]_{|z=1}$. Then*

$$\frac{d}{dt} \mathrm{E}\, f(N(t)) = \mathrm{E}\, f_H(N(t)) .$$

Proof. Start from the Schrödinger equation analog $\frac{\partial}{\partial t}\phi(t,z) = H\,\phi(t,z)$. The partial derivation $\partial/\partial t$ commutes with the operator $f(\theta)$ and the evaluation at $z = 1$. Thus $\partial/\partial t\, f(\theta)\,\phi(t,z)_{|_{z=1}} = f(\theta)\,H\,\phi(t,z)_{|_{z=1}}$. The left-hand side of the equality is equal to $d/dt\, \mathrm{E}\,f(N(t))$ according to Lemma 2. In the right-hand side, the product $f(\theta)H$ van be replaced by the commutator $[f(\theta), H]$ since, according to Formula (11), the Hamiltonian H is zero at $z = 1$. The second member is equal to $\mathrm{E}\,f_H(N(t))$ by Lemma 2.

The Algorithm

Input: *A stochastic Petri net $\mathcal{R}$ and a maximum order $q \in \mathbb{N}$.*
Output: *A linear differential system characterizing the time evolution of moments, up to degree q.*

1. *Compute the Hamiltonian H of $\mathcal{R}$ using formula (11).*
2. *For all multi-indices $\kappa \in \mathbb{N}^n$ such that $|\kappa| \leq q$, compute the commutator evaluated at $z = 1$*

$$f_\kappa = [\theta^\kappa, H]_{|_{z=1}}, \qquad (f_\kappa \in \mathbb{R}[\theta_1, \ldots, \theta_n])\,, \tag{12}$$

Then generate the linear differential equation $\dfrac{d}{dt}\,\mathrm{E}\,N^\kappa(t) = \mathrm{E}\,f_\kappa(N(t))$, using Proposition 1.

The polynomial f_κ in Formula (12) can be computed in different ways. It can be computed by using a smart computation of the Lie bracket (which avoids the expansion of the Lie bracket by computing $\theta^\kappa H$ and then substracting $H\,\theta^\kappa$) and a specialization to $z = 1$. Otherwise, it is possible to compute $f_\kappa = \theta^\kappa\,H_{|_{z=1}}$ since the term $H\theta^\kappa_{|_{z=1}}$ cancels.

Remark 1. The algorithm is stated using Weyl algebras computations since it makes the proofs easier to establish. Section 6 also provides a formula only based on basic operations on commutative polynomials in θ. As detailed in Section 6, we believe that avoiding Weyl algebras computations is easier from a software implementation point of view.

The returned ODE system is truncated. Thus, some ODE may depend on moments of order higher than q, for which no ODE is generated. This problem does not occur in Example (8). The Hamiltonian is

$$H = \lambda(z-1) + \mu\left(\frac{1}{z} - 1\right)\theta \quad \text{with} \quad \theta = z\frac{\partial}{\partial z}\,.$$

The algorithm computes the brackets

$$[\theta, H]_{|_{z=1}} = \lambda - \mu\theta\,, \qquad [\theta^2, H]_{|_{z=1}} = \lambda + (2\lambda + \mu)\,\theta - 2\mu\theta^2\,.$$

and returns the differential system

$$\frac{d}{dt}\,\mathrm{E}\,N(t) = \lambda - \mu\,\mathrm{E}\,N(t)\,, \qquad \frac{d}{dt}\,\mathrm{E}\,N^2(t) = \lambda + (2\lambda + \mu)\,\mathrm{E}\,N(t) - 2\mu\,\mathrm{E}\,N^2(t)\,.$$

The variance is computed using $\operatorname{Var} N(t) = \operatorname{E} N^2(t) - (\operatorname{E} N(t))^2$. The dynamics of the mean $x(t) = \operatorname{E} N(t)$ and the variance $v(t) = \operatorname{Var} N(t)$ follows immediately.

$$\frac{d}{dt}x(t) = \lambda - \mu x(t) , \quad \frac{d}{dt}v(t) = \lambda + \mu x(t) - 2\mu v(t) .$$

5 Model Reduction, Model Restriction and Conservation Laws

Definition 5. *Let λ be a n-dimensional vector of integers. A conservation law I_λ is a linear combination of the following form, with integer coefficients, which is conserved by each transition of the considered Petri net:*

$$I_\lambda(\nu) = \lambda_1\nu_1 + \lambda_2\nu_2 + \cdots + \lambda_n\nu_n, \qquad (\nu \in \mathbb{N}^n, \ \lambda \in \mathbb{Z}^n).$$

This notion is independent of the temporisation, hence of the kinetic constants associated to chemical reactions. Recall that the incidence matrix of a Petri net is the transpose of the stoichiometry matrix of a chemical reaction system. The next lemma is then well-known [9, sect. 5.3].

Lemma 3. *Let $\mathcal{R}$ be a Petri net. The column vector $\lambda = (\lambda_1, \lambda_2, \ldots, \lambda_n) \in \mathbb{Z}^n$ defines a conservation law iff $C\lambda = 0$ where $C = C^+ - C^-$ is the incidence matrix of $\mathcal{R}$.*

5.1 Model Reduction

A conservation law I_λ induces a graduation w_λ of the algebra $\mathrm{Weyl}_\mathbb{R}(z)$ defined by $\mathrm{w}_\lambda(\partial/\partial z_k) = -\lambda_k$ and $\mathrm{w}_\lambda(z_k) = \lambda_k$ for all $k = 1 \ldots n$. Define, moreover, $\mathrm{w}_\lambda(z^\nu) = \sum_k \lambda_k \nu_k$, that is $\mathrm{w}_\lambda(z^\nu) = I_\lambda(\nu)$, for all $\nu \in \mathbb{N}^n$.

Lemma 4. *Assume that a conservation law I_λ holds. Then, $\phi(t, z)$ is a formal power series, homogeneous for w_λ, with weight $\mathrm{w}_\lambda(\phi(t, z)) = I_\lambda(\nu_0)$, where the multi-index $\nu_0 \in \mathbb{N}^n$ denotes the initial state at $t = 0$.*

Proof. As $I_\lambda(\nu)$ is independent of the time t, the fact that $I_\lambda(\nu) \neq I_\lambda(\nu_0)$ implies that $\pi_\nu(t) = 0$ for any $t \geq 0$. As a consequence, the series $\phi(t, z) = \sum_\nu \pi_\nu(t) z^\nu$ is w_λ-homogeneous of weight $\mathrm{w}_\lambda(z^{\nu_0})$.

Denote $C_0 = I_\lambda(\nu_0) \in \mathbb{Z}$. Then the operator $I_\lambda(\theta) - C_0 = \sum_k \lambda_k \theta_k - C_0$ vanishes on the generating series $\phi(t, z)$ because, according to Lemma 4, one has $I_\lambda(\theta)\phi(t, z) = C_0\phi(t, z)$. Every operator of the left ideal generated in $\mathrm{Weyl}_\mathbb{R}(z)$ by $I_\lambda(\theta) - C_0$ has therefore a null action on $\phi(t, z)$.

The Model Reduction Algorithm
Input: A Hamiltonian $H \in \mathrm{Weyl}_\mathbb{R}(z_1, \ldots, z_n)$ describing the evolution of the multi-index $N(t) = (N_1(t), \ldots, N_n(t))$ and a conservation law I_λ. One assumes, without loss of generality, that λ_n is nonzero.

Output: A reduced Hamiltonian $H' \in \text{Weyl}_{\mathbb{R}}(z_1, \ldots, z_{n-1})$ describing the evolution of $(N_1(t), \ldots, N_{n-1}(t))$.
The new Hamiltonian H' is obtained from H by the following substitution

$$\theta_n \mapsto \frac{1}{\lambda_n} \left[C_0 - (\lambda_1 \theta_1 + \lambda_2 \theta_2 + \cdots + \lambda_{n-1} \theta_{n-1}) \right] , \qquad z_n \mapsto 1 .$$

5.2   Model Restriction

The presence of conservation laws often enables us to bound some random variables of the model. Those bounds in general depend on the initial state $\nu_0 \in \mathbb{N}^n$. In this situation, it is possible to restrict the model by taking a quotient of the Weyl algebra $\text{Weyl}_{\mathbb{R}}(z_1, \ldots, z_n)$ by a left ideal. The method is presented over the following system of chemical reactions

$$R_1 + R_2 \xrightarrow{\lambda} R_3 , \qquad R_3 \xrightarrow{\mu} R_1 + R_2 ,$$

together with initial conditions $\nu^0 = (a, b, 0)$ for $a, b \in \mathbb{N}$. The Petri net $\mathcal{R}$ admits two conservation laws $\nu_1 + \nu_3 = a$ and $\nu_2 + \nu_3 = b$. It is therefore possible to consider the random variable $N_3(t)$ only, with the bound $0 \leq N_3(t) \leq \min(a, b)$. Our software computes the Hamiltonian:

$$H = \lambda \left(\frac{z_3}{z_1 z_2} - 1 \right) \theta_1 \theta_2 + \mu \left(\frac{z_1 z_2}{z_3} - 1 \right) \theta_3 .$$

Performing the substitutions $\theta_1 \mapsto a - \theta_3$, $\theta_2 \mapsto b - \theta_3$, $z_1 \mapsto 1$ and $z_2 \mapsto 1$, leads, to the reduced Hamiltonian $H' = \lambda (z_3 - 1)(b - \theta_3)(b - \theta_3) + \mu (z_3^{-1} - 1)\theta_3$, which is an element of $\text{Weyl}_{\mathbb{R}}(z_3)$. The bound $N_3(t) \leq m$ with $m = \min(a, b)$ implies that the function $f(\nu_3) = \nu_3(\nu_3 - 1) \cdots (\nu_3 - m)$ is zero at all time t. One then gets

$$f(\theta_3) \, \phi(t, z_3) = \sum_{0 \leq \nu_3 \leq m} \pi_{\nu_3}(t) \, f(\nu_3) \, z_3^{\nu_3} = 0, \quad \forall t \in \mathbb{R}_{\geq 0} .$$

The operator $f(\theta_3)$ vanishes on the generating series $\phi(t, z_3)$. All computations can therefore be done in the quotient of the algebra $\text{Weyl}_{\mathbb{R}}(z_3)$ by the left ideal generated by $f(\theta_3)$. From an algorithmic point of view, this can be achieved by Gröbner basis techniques on the polynomials f_κ (which are commutative polynomials in θ).

6   A Combined Formula for Differentiating and Evaluating

In this section, the explicit Formula (14) is given for computing f_κ (Formula 12) which are commutators evaluated at $z = 1$. We believe that this formula has several advantages. First, this combined formula provides an improvement of the algorithm of Section 4. Indeed, Formula (14) can be implemented only with basic operations on commutative polynomials in θ. Moreover, Formula (14) does not need any specialization $z = 1$. Second, all computations in the Weyl algebra can be avoided, which makes the algorithm easier to implement since no library for computing in Weyl algebras is needed. Indeed, Weyl algebras computations are, to our knowledge, only available through softwares like Maple [2,1], Macaulay 2 [10], which makes it hard to produce an independant or GPL standalone library. Third, we believe that Formula (14) helps finding interesting formulas such as in Lemma 9 (page 94). Fourth, the computation of Formula (14) can be easily mixed with the reductions detailed in section 5 by using modular exponentiations.

Lemma 5. *Let A and B taken in $\mathrm{Weyl}_{\mathbb{R}}(z_1, \ldots, z_n)$. Define $\mathrm{ad}_A(B) = [A, B]$. Then*

$$A^k B = \sum_{i=0}^{k} \binom{k}{i} \mathrm{ad}_A^i(B) A^{k-i} \ , \tag{13}$$

where ad_A^i denotes the ad_A function composed i times.

Proof. Let t be an indeterminate. The classical identity, between formal series in t, $\exp(tA) B \exp(-tA) = \exp(t \, \mathrm{ad}_A)(B)$ holds (adjoint representation of a Lie group over its Lie algebra, see [3] for details). This formula can be rewritten as $\exp(tA) B = \exp(t \, \mathrm{ad}_A)(B) \exp(tA)$ and developed with a Taylor expansion. The result is proved by identifying the coefficients of t^k in each side of

$$\sum_{k \geq 0} \frac{t^k}{k!} A^k B = \sum_{i,j \geq 0} \frac{t^{i+j}}{i! \, j!} \, \mathrm{ad}_A^i(B) A^j \ .$$

Lemma 6. *For all $m \in \mathbb{Z}$ and any $k \in \mathbb{N}$, one has $\theta^k z^m = z^m (m + \theta)^k$ with $\theta = z \frac{\partial}{\partial z}$.*

Proof. One has $\mathrm{ad}_\theta(z^m) = [\theta, z^m] = m z^m$. Thus, $\mathrm{ad}_\theta^i(z^m) = m^i z^m$ for any $i \geq 0$. The lemma then follows from Lemma 5 by taking $A = \theta$ and $B = z^m$.

Let $\nu \in \mathbb{Z}^n$ and $\kappa \in \mathbb{N}^n$. Denote $(\nu + \theta)^\kappa = (\nu_1 + \theta_1)^{\kappa_1} (\nu_2 + \theta_2)^{\kappa_2} \cdots (\nu_n + \theta_n)^{\kappa_n}$.

Lemma 7. *The commutation relation between θ^κ and z^ν, viewed as an element of $\mathrm{Weyl}_{\mathbb{R}}(z_1, z_2, \ldots, z_n)$, can now be written:*

$$\theta^\kappa z^\nu = z^\nu (\nu + \theta)^\kappa \ .$$

Proof. The proof relies on Lemma 6. It is only given for $n = 3$.

$$\begin{aligned}
\theta^\kappa z^\nu &= (\theta_1^{\kappa_1} \theta_2^{\kappa_2} \theta_3^{\kappa_3})(z_1^{\nu_1} z_2^{\nu_2} z_3^{\nu_3}) = (\theta_1^{\kappa_1} z_1^{\nu_1})(\theta_2^{\kappa_2} z_2^{\nu_2})(\theta_3^{\kappa_3} z_3^{\nu_3}) \\
&= z_1^{\nu_1}(\nu_1 + \theta_1)^{\kappa_1} z_2^{\nu_2}(\nu_2 + \theta_2)^{\kappa_2} z_3^{\nu_3}(\nu_3 + \theta_3)^{\kappa_3} \\
&= (z_1^{\nu_1} z_2^{\nu_2} z_3^{\nu_3})(\nu_1 + \theta_1)^{\kappa_1}(\nu_2 + \theta_2)^{\kappa_2}(\nu_3 + \theta_3)^{\kappa_3} = z^\nu (\nu + \theta)^\kappa \ .
\end{aligned}$$

Proposition 2. *Let $\mathcal{R}$ be a stochastic Petri net, the Hamiltonian of which is H. Then*

$$[\theta^\kappa, H]_{|_{z=1}} = \sum_{(c,\alpha,\beta) \in \mathcal{R}} \frac{c}{\alpha!} \left[(\beta - \alpha + \theta)^\kappa - \theta^\kappa \right] \theta^\alpha \ , \qquad (\kappa \in \mathbb{N}^n) \ . \tag{14}$$

Proof. The evaluated commutator $[\theta^{\kappa}, H]_{|_{z=1}}$ is linear in H. For simplicity, it is assumed that a single chemical reaction is involved, so that $H = (z^{\beta-\alpha} - 1)\,\theta^{\underline{\alpha}}$. The proof reduces to the computation of $[\theta^{\kappa}, (z^{\beta-\alpha}-1)\,\theta^{\underline{\alpha}}]_{|_{z=1}}$. Lemma 7 is used to make the terms θ^{κ} and $z^{\beta-\alpha} - 1$ commute. Computations give the following formula

$$\theta^{\kappa}\,(z^{\beta-\alpha} - 1) = \theta^{\kappa}\,z^{\beta-\alpha} - \theta^{\kappa} = z^{\beta-\alpha}(\beta - \alpha + \theta)^{\kappa} - \theta^{\kappa} \ .$$

Evaluate it at $z = 1$ and apply the fact that $H_{|_{z=1}} = 0$. The proposition is proved.

7   Order of a Chemical Reaction System

In this section, well-known results on order 1 systems are recovered from the Weyl algebra theory (see [12] and [4, sect. 5.3.3]). First, some further theoretical developments on Weyl algebra are introduced.

The algebra $\mathrm{Weyl}_{\mathbb{R}}(z) = \mathrm{Weyl}_{\mathbb{R}}(z_1, \ldots, z_n)$ is graded by the *degree*. It is readily checked that the product AB of two elements $A, B \in \mathrm{Weyl}_{\mathbb{R}}(z)$ homogeneous by degree, is also homogeneous of degree $\deg(A) + \deg(B)$. Moreover, the algebra is filtered by the *order* (in the sense of differential operators). The component $\mathcal{F}_q \subset \mathrm{Weyl}_{\mathbb{R}}(z)$, $q \in \mathbb{N}$, of the growing filtration

$$\mathcal{F}: \ \mathcal{F}_0 \subset \mathcal{F}_1 \subset \mathcal{F}_2 \subset \cdots \tag{15}$$

is the $\mathbb{R}$-vector space spanned by the elements $z^{\alpha}\partial_z^{\beta}$ of order *at most* q, i.e. such that $|\beta| \leq q$. It is possible to check that $\mathcal{F}_k \mathcal{F}_l \subset \mathcal{F}_{k+l}$ for all $k, l \in \mathbb{N}$. The *graded* algebra associated to the filtration $\mathcal{F}$ is commutative since $[\mathcal{F}_k, \mathcal{F}_l] \subset \mathcal{F}_{k+l-1}$ for all $k, l \in \mathbb{N}$ (with the convention $\mathcal{F}_{-1} = \{0\}$). It is defined by

$$\mathrm{gr}\,\mathrm{Weyl}_{\mathbb{R}}(z) = \mathcal{F}_0 \oplus \mathcal{F}_1/\mathcal{F}_0 \oplus \mathcal{F}_2/\mathcal{F}_1 \oplus \cdots \tag{16}$$

Definition 6. *A chemical reaction system $\mathcal{R}$ is said to be of* order $q \in \mathbb{N}$, *if each reaction $(c, \alpha, \beta) \in \mathcal{R}$ satisfies $|\alpha| \leq q$ (where $|\alpha| = \alpha_1 + \alpha_2 + \cdots + \alpha_n$ for any $\alpha \in \mathbb{N}^n$).*

Thus, a chemical reaction system $\mathcal{R}$ is of order q if every reaction of $\mathcal{R}$ consumes at most q molecules. The next lemma follows immediately from Formula (11).

Lemma 8. *A chemical reaction system is of order q iff its Hamiltonian H belongs to the $\mathcal{F}_q$ component of the Weyl algebra filtration defined by (15).*

Any polynomial $f \in \mathbb{R}[\theta_1, \theta_2, \ldots, \theta_n]$, homogeneous of degree d, defines a degree d moment $\mathrm{E}\,f(N(t))$.

Proposition 3. *In a chemical reaction system of order q, the derivative with respect to the time of a degree d moment, only depends on other moments of order at most $q + d - 1$.*

Proof. Let $H \in \mathrm{Weyl}_{\mathbb{R}}(z_1, z_2, \ldots, z_n)$ be an order q operator i.e. $H \in \mathcal{F}_q$. If $f \in \mathbb{R}[\theta_1, \theta_2, \ldots, \theta_n]$ is an homogeneous polynomial of degree d, then one has $[f(\theta), H] \in \mathcal{F}_{q+d-1}$. Evaluation at $z = 1$ yields $[f(\theta), H]_{|_{z=1}} \in \mathbb{R}[\theta_1, \theta_2, \ldots, \theta_n]$ of degree at most $q + d - 1$.

Corollary 1. *In a chemical reaction system, the dynamics of the mean values* $\mathrm{E}\, N_k(t)$, $k = 1 \ldots n$, *(i.e. first order moments) can be written as follows (with* $k \in \mathbb{N}$, $\alpha, \beta \in \mathbb{N}^n$*). Moreover, the dynamics of a first order system is linear in the variables* $\mathrm{E}\, N_k(t)$.

$$\frac{d}{dt} \mathrm{E}\, N_k(t) = \sum_{(c,\alpha,\beta) \in \mathcal{R}} \frac{c}{\alpha!} (\beta_k - \alpha_k)\, \mathrm{E}\, N(t)^{\underline{\alpha}}$$

Proof. One deduces $\theta_k H = \sum_{(c,\alpha,\beta) \in \mathcal{R}} \frac{c}{\alpha!} (\beta_k - \alpha_k) z^{\beta-\alpha} \theta^{\underline{\alpha}} + \frac{c}{\alpha!} (z^{\beta-\alpha} - 1) \theta_k \theta^{\underline{\alpha}}$ from Formula (11). The evaluation at the point $z = 1$ leads to the relation $\theta_k H_{|_{z=1}} = \sum_{(c,\alpha,\beta) \in \mathcal{R}} \frac{c}{\alpha!} (\beta_k - \alpha_k) \theta^{\underline{\alpha}}$. Applying Lemma 2 and Proposition 1, the formula is proved. Whenever $|\alpha| \leq 1$, $\alpha! = 1$ and $N(t)^{\underline{\alpha}} = N(t)^{\alpha}$. Moreover $N(t)^{\alpha} = 1$ iff $\alpha = 0$ and $N(t)^{\alpha} = N_j(t)$ if $\alpha = (0, \ldots, 0, 1, 0, \ldots, 0)$, the "1" occuring in j^{th} position.

The deterministic models in chemical kinetics, classically built using the mass action law, are recovered. For first order systems, the deterministic model corresponds to an unbiased averaging of the random variables $N_k(t)$. It does not apply to covariance matrices $\mathrm{Cov}(N_i(t), N_j(t))$, where $i, j = 1 \ldots n$.

8   Examples

8.1   First Order Systems

The following model of a non-regulated gene [11] is interesting because it was verified experimentally and because it shows the importance of random phenomena in gene expression. The transcription of the gene produces messenger RNAs (mRNA) which are translated into proteins:

$$\begin{cases} \mathrm{DNA} \xrightarrow{k_R} \mathrm{DNA} + \mathrm{mRNA} & \text{(transcription)} \\ \mathrm{mRNA} \xrightarrow{k_P} \mathrm{mRNA} + \mathrm{protein} & \text{(translation)} \\ \mathrm{mRNA} \xrightarrow{\gamma_R} \emptyset & \text{(mRNA degradation)} \\ \mathrm{protein} \xrightarrow{\gamma_P} \emptyset & \text{(protein degradation)} \end{cases} \qquad (17)$$

The three chemical species are numbered is the following manner: $R_1 = \mathrm{DNA}$, $R_2 = \mathrm{mRNA}$ and $R_3 = \mathrm{protein}$. It admits the conservation law $\nu_1 = 1$, meaning that there is only one gene involved in the network (at any time t). The model is reduced by setting $\theta_1 = 1$ and $z_1 = 1$. The reduced Hamiltonian is equal to

$$H = k_R (z_2 - 1) + k_P (z_3 - 1) \theta_2 + \gamma_R (z_2^{-1} - 1) \theta_2 + \gamma_P (z_3^{-1} - 1) \theta_3$$

Only two state variables remain $\nu = (\nu_2, \nu_3) \in \mathbb{N}^2$, the number of mRNA and the number of proteins at time t. Let $N(t) = (N_2(t), N_3(t))$ be the stochastic process associated to (17). We are going to compute the means $x_2(t) = \mathrm{E}\, N_2(t)$ et $x_3(t) = \mathrm{E}\, N_3(t)$ and the covariance matrix $x_{ij}(t) = \mathrm{Cov}(N_i(t), N_j(t))$ for $i, j = 2, 3$. According to the algorithm of Section 4, our software computes the commutators evaluated at the point $z_2 = z_3 = 1$:

$$[\theta_2, H]\big|_{|z=1} = k_R - \gamma_R \theta_2$$

$$[\theta_3, H]\big|_{|z=1} = k_P \theta_2 - \gamma_P \theta_3$$

$$[\theta_2^2, H]\big|_{|z=1} = k_R + (2k_R + \gamma_R)\theta_2 - 2\gamma_R \theta_2^2$$

$$[\theta_2\theta_3, H]\big|_{|z=1} = k_R \theta_3 + k_P \theta_2^2 + (-\gamma_R - \gamma_P)\theta_2\theta_3$$

$$[\theta_3^2, H]\big|_{|z=1} = k_P \theta_2 + \gamma_P \theta_3 + 2k_P \theta_2\theta_3 - 2\gamma_P \theta_3^2$$

and generates the system which describes the time evolution of the means and the covariances:

$$\frac{d}{dt}x_2(t) = k_R - \gamma_R\, x_2(t)$$

$$\frac{d}{dt}x_3(t) = k_P\, x_2(t) - \gamma_P\, x_3(t)$$

$$\frac{d}{dt}x_{2,2}(t) = \gamma_R\, x_2(t) + k_R - 2\,\gamma_R\, x_{2,2}(t) \qquad (18)$$

$$\frac{d}{dt}x_{2,3}(t) = (-\gamma_R - \gamma_P)\, x_{2,3}(t) + k_P\, x_{2,2}(t)$$

$$\frac{d}{dt}x_{3,3}(t) = k_P\, x_2(t) - 2\,\gamma_P\, x_{3,3}(t) + 2\,k_P\, x_{2,3}(t) + \gamma_P\, x_3(t)$$

Since the chemical reaction system has order 1, a linear system is obtained and the phenomenon of the infinite cascade does not occur. For simplification, a time scale such that $\gamma_R = 1$ is selected. The computation of the means and the variances at the stationary state give:

$$x_2 = k_R, \qquad x_3 = \frac{k_P k_R}{\gamma_P},$$

$$x_{2,2} = k_R, \qquad x_{2,3} = \frac{k_P k_R}{1 + \gamma_P}, \qquad x_{3,3} = \frac{k_P k_R (\gamma_P + k_P + 1)}{\gamma_P (1 + \gamma_P)}$$

This result is exact. The same formulas appear in [11] and are proved by using Langevin's technique: the two first equations of (18) are viewed as a deterministic model (arising from the mass action law). Then, two white noises of zero average are incorporated in the righthand sides. This method is difficult to justify theoretically. By solving the equations (18), by, say, a Laplace transform technique, we obtain exact formulas for the means and the variances during the transient stage, given below:

$$x_2(t) = k_R \left(-e^{-t} + 1\right)$$
$$x_3(t) = \frac{k_P k_R \left(-1 + e^{-\gamma_P t} + \gamma_P \left(-e^{-t} + 1\right)\right)}{\gamma_P \left(-1 + \gamma_P\right)}$$
$$x_{2,2}(t) = k_R \left(-e^{-t} + 1\right)$$
$$x_{2,3}(t) = \frac{\left(-e^{-t} \left(1 + \gamma_P\right) + \gamma_P + e^{-(1+\gamma_P)t}\right) k_R k_P}{\gamma_P \left(1 + \gamma_P\right)}$$

8.2   Second Order Systems

The dynamics of the degree d moments depends on moments of degree strictly greater than d. In other words, an infinite cascade occurs. This is the main source of difficulty. In order to break an infinite cascade there are two possible methods:

1. One can operate an approximation assuming that the centered moments of degree d are zero for d large enough. This is a legitimate approximation whenever the number of tokens in each place (the number of chemical molecules of each chemical species) remains high at any time t.
2. One has a relation expressing degree $d + 1$ moments as a function of moments of degree at most d. This case occurs, in particular, whenever random variables only take a finite number of different values.

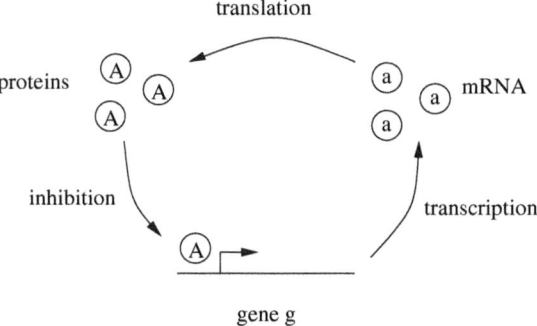

Fig. 3. Autoregulated gene

Autoregulated Gene. The transcription of the gene produces messenger RNA, which in turn are translated in proteins. If a protein binds to the gene, then the transcription is blocked. (see Figure 3).

$$\text{gene} \xrightarrow{\lambda_1} \text{gene} + \text{mRNA}$$
$$\text{mRNA} \xrightarrow{\lambda_2} \text{mRNA} + \text{protein}$$
$$\text{mRNA} \xrightarrow{\mu_1} \emptyset$$
$$\text{protein} \xrightarrow{\mu_2} \emptyset$$
$$\text{gene} + \text{protein} \xrightarrow{c_1} \text{blocked\_gene}$$
$$\text{blocked\_gene} \xrightarrow{c_2} \text{gene} + \text{protein}$$

The four chemical species are numbered: $R_1 = $ mRNA, $R_2 = $ gene, $R_3 = $ blocked gene, $R_4 = $ protein. That system obeys the conservation law $I(\nu) = \nu_2 + \nu_3$, meaning that the total number of "molecules" of type *gene* and *blocked_gene* remains constant. In practice, only one gene is involved. Therefore the Petri net gets initialized with the following assumption $\nu_2 + \nu_3 = 1$. Since at any time t, $\nu_2 + \nu_3 = 1$, we choose to remove the state variable ν_3 putting $\nu_3 = 1 - \nu_2$. After model reduction, the Hamiltonian becomes

$$H = \lambda_1 (z_1 - 1) \theta_2 + \lambda_2 (z_4 - 1) \theta_1 + \mu_1 \left(\frac{1}{z_1} - 1 \right) \theta_1 + \mu_2 \left(\frac{1}{z_4} - 1 \right) \theta_4$$
$$+ c_1 \left(\frac{1}{z_2 z_4} - 1 \right) \theta_2 \theta_4 + c_2 (z_2 z_4 - 1)(1 - \theta_2) .$$

We introduce the means $x_i(t) = \mathrm{E}\, N_i(t)$ for $i = 1, 2, 4$ and the covariance matrix $x_{ij}(t) = \mathrm{Cov}(N_i(t), N_j(t))$ for $i, j = 1, 2, 4$. The random variable $N_2(t)$ is boolean $(\nu_2 = \nu_2^2)$, the model restriction algorithm can therefore be applied. The algorithm computes the brackets $[\theta^\kappa, H]$ in the Weyl algebra quotiented by the left ideal spanned by the relation $\theta_2 = \theta_2^2$. The evaluation at $z = 1$ is performed afterwards. It yields:

$$[\theta_1, H]\big|_{z=1} = -\mu_1 \theta_1 + \lambda_1 \theta_2$$
$$[\theta_2, H]\big|_{z=1} = c_2 - c_2 \theta_2 - c_1 \theta_2 \theta_4$$
$$[\theta_4, H]\big|_{z=1} = c_2 + \lambda_2 \theta_1 - c_2 \theta_2 - \mu_2 \theta_4 - c_1 \theta_2 \theta_4$$
$$[\theta_1^2, H]\big|_{z=1} = \mu_1 \theta_1 + \lambda_1 \theta_2 - 2\mu_1 \theta_1^2 + 2\lambda_1 \theta_1 \theta_2$$
$$[\theta_2 \theta_1, H]\big|_{z=1} = c_2 \theta_1 + \lambda_1 \theta_2 - (\mu_1 + c_2) \theta_1 \theta_2 - c_1 \theta_1 \theta_2 \theta_4$$

An infinite cascade occurs because the dynamics on the degree 2 moments involves two degree three moments, namely the moment coded by the following operators $\theta_1 \theta_2 \theta_4$ and $\theta_2 \theta_4^2$. The approximation assumption that all centered moments of degree 3 are zero, can be used to break this cascade. Consider three random variables (X_1, X_2, X_3) with respective means (x_1, x_2, x_3). A routine computation gives the centered moment of order three $\mathrm{E}((X_1 - x_1)(X_2 - x_2)(X_3 - x_3))$ as $\mathrm{E}(X_1 X_2 X_3) - x_1 x_2 x_3 - x_1 \mathrm{Cov}(X_2, X_3) - x_2 \mathrm{Cov}(X_3, X_1) - x_3 \mathrm{Cov}(X_1, X_2)$. We have no proof that this approximation is best fit. Our software computes the ordinary *non-linear* differential equation system:

$$\frac{d}{dt} x_1(t) = -\mu_1 x_1(t) + \lambda_1 x_2(t)$$
$$\frac{d}{dt} x_2(t) = c_2 - c_2 x_2(t) - c_1 x_4(t) x_2(t) - c_1 x_{2,4}(t)$$
$$\frac{d}{dt} x_4(t) = c_2 + \lambda_2 x_1(t) - c_2 x_2(t) - c_1 x_4(t) x_2(t)$$
$$\qquad - \mu_2 x_4(t) - c_1 x_{2,4}(t)$$

$$\frac{d}{dt}x_{1,1}(t) = \mu_1 x_1(t) + 2\lambda_1 x_{1,2}(t) + \lambda_1 x_2(t) - 2\mu_1 x_{1,1}(t)$$
$$\frac{d}{dt}x_{1,2}(t) = \lambda_1 x_2(t) - c_1 x_{1,4}(t) x_2(t) - c_1 x_4(t) x_{1,2}(t)$$
$$- (\mu_1 + c_2)x_{1,2}(t) - \lambda_1 x_2(t)^2$$

The Figures 4, 5 and 6 show the numerical simulations produced by our software. We used the parameter values $\lambda_1 = 30.0$, $\lambda_2 = 10.0$, $\mu_1 = \mu_2 = 0.1$, and $c_1 = c_2 = 1.0$, and the initial conditions $x_1(0) = 0$ (mRNA), $x_2(0) = 1$ (gene), $x_3(0) = 0$ (protein).

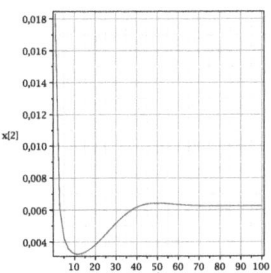
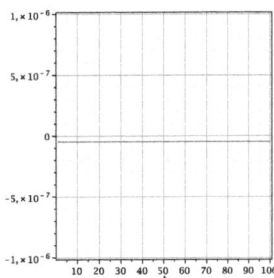

(a) Average expression rate of the gene

(b) Accuracy test on variance

Fig. 4. Autoregulated gene (boolean variable). Simulations are consistent with the fact that any boolean variable X satifies $\mathrm{Var}\,X = x(1-x)$, where $x = \mathrm{E}\,X$.

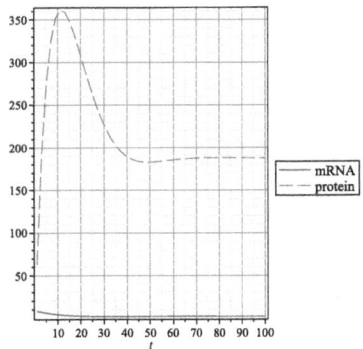
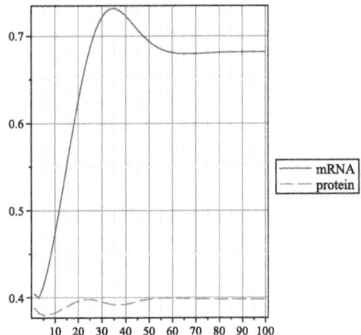

(a) Average number of mRNA and proteins

(b) Relative standard deviation

Fig. 5. Joint evolution of mRNA and protein number. The relative standard deviation of a random variable X is σ/x with $\sigma^2 := \mathrm{Var}\,X$ and $x := \mathrm{E}(X)$.

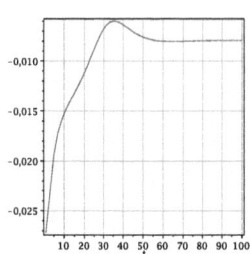 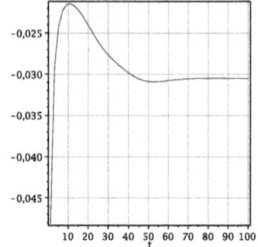 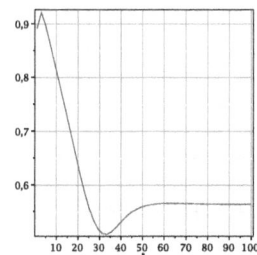

(a) Gene-mRNA corre-   (b) Gene-protein corre-   (c) mRNA-protein cor-
lation   lation   relation

Fig. 6. Evolution over time of correlation rates. The correlation rate $c(X, Y)$ between two random variables X and Y is defined by $c(X, Y) := \frac{\text{Cov}(X,Y)}{\sigma(X)\,\sigma(Y)}$.

Second Order Degradation. Consider the chemical reaction

$$2R \xrightarrow{\mu} \emptyset \tag{19}$$

Using a time dilatation, it is assumed that $\mu = 1$. The Hamiltonian is then, with $\theta = z\,\partial/\partial z$:

$$H = \frac{1}{2}\left(\frac{1}{z^2} - 1\right)\theta(\theta - 1) = \frac{1}{2}(1 - z^2)\left(\frac{\partial}{\partial z}\right)^2 .$$

On this example, the dynamics on the moments $E\,N(t)^k$, $k = 1\ldots 4$ is coded by

$$\begin{aligned}
[\theta, H]_{|z=1} &= -\theta^2 + \theta \\
[\theta^2, H]_{|z=1} &= -2\,\theta^3 + 4\,\theta^2 - 2\,\theta \\
[\theta^3, H]_{|z=1} &= -3\,\theta^4 + 9\,\theta^3 - 10\,\theta^2 + 4\,\theta \\
[\theta^4, H]_{|z=1} &= -4\,\theta^5 + 16\,\theta^4 - 28\,\theta^3 + 24\,\theta^2 - 8\,\theta
\end{aligned} \tag{20}$$

Lemma 9. *For all $m \in \mathbb{N}$, Formulas (20) take the following closed form:*

$$[\theta^m, H]_{|z=1} = \frac{1}{2}\big[(-2 + \theta)^m - \theta^m\big]\theta(\theta - 1) \tag{21}$$

Proof. Apply Formula (14) for $(c, \alpha, \beta) := (1, (2), (0))$.

Let $x(t)$ denote the mean and $v(t)$ the variance of $N(t)$. The previously developed approximation, obtained by killing centered order 3 moments, then gives

$$\frac{d}{dt}x(t) = x(t) - v(t) - (x(t))^2$$

$$\frac{d}{dt}v(t) = -2\,x(t) + 4\,v(t) + 2\,(x(t))^2 - 4\,v(t)\,x(t) .$$

This approximation method behaves badly when $E\,N(t)$ gets smaller than 1 (see Figure 7). Unfortunately the situation does not improve if the centered moment is kept up to a higher order.

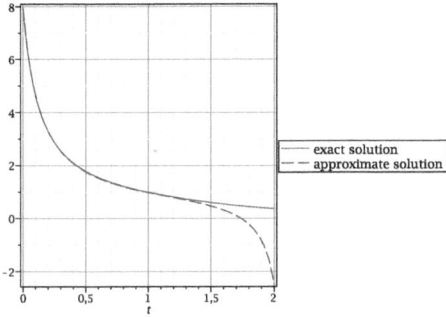

Fig. 7. Evolution of the mean $x(t) = \mathrm{E}\,N(t)$ over time, starting with state $n_0 = 8$. The approximation gets worse after $x(t)$ crosses the value 1 (around $t = 1$).

We show now how to get the *exact* dynamics on the moments of any order. Assume, as an example, that the initial state is $n_0 = 8$. Then, the polynomial function $f(\nu) = \nu(\nu - 2)(\nu - 4)(\nu - 6)(\nu - 8)$ is zero at any instant t. According to Lemma 2 and Proposition 1, we have to quotient the algebra $\mathrm{Weyl}_\mathbb{R}(z)$ by the left ideal spanned by the relation $f(\theta) = 0$. This leads to $f(\theta) = \theta\,(\theta - 2)\,(\theta - 4)\,(\theta - 6)\,(\theta - 8)$. Adding this extra relation to System (20) enables us to get order 5 moments as functions of the moments of order at most 4. This way, we get the exact dynamics, a linear system, describing the time evolution of the moments $x_k(t) = \mathrm{E}\,N^k(t)$, for $k = 1 \ldots 4$.

$$\frac{d}{dt}x_1\,(t) = x_1\,(t) - x_2\,(t)$$

$$\frac{d}{dt}x_2\,(t) = -2\,x_1\,(t) + 4\,x_2\,(t) - 2\,x_3\,(t)$$

$$\frac{d}{dt}x_3\,(t) = 4\,x_1\,(t) - 10\,x_2\,(t) + 9\,x_3\,(t) - 3\,x_4\,(t)$$

$$\frac{d}{dt}x_4\,(t) = 1528\,x_1\,(t) - 1576\,x_2\,(t) + 532\,x_3\,(t) - 64\,x_4\,(t)$$

with the initial conditions: $x_1(0) = 8$, $x_2\,(0) = 64$, $x_3\,(0) = 512$, $x_4\,(0) = 4096$. These linear differential equations get solved by means of the Laplace transform:

$$x_1(t) = \frac{8}{3}\,\mathrm{e}^{-t} + \frac{112}{33}\,\mathrm{e}^{-6\,t} + \frac{64}{39}\,\mathrm{e}^{-15\,t} + \frac{128}{429}\,\mathrm{e}^{-28\,t}$$

$$x_2(t) = \frac{16}{3}\,\mathrm{e}^{-t} + \frac{784}{33}\,\mathrm{e}^{-6\,t} + \frac{1024}{39}\,\mathrm{e}^{-15\,t} + \frac{3712}{429}\,\mathrm{e}^{-28\,t} \tag{22}$$

$$x_3(t) = \cdots$$

The infinite cascade if thereby broken by an exact method (see the simulations shown in Figure 8)

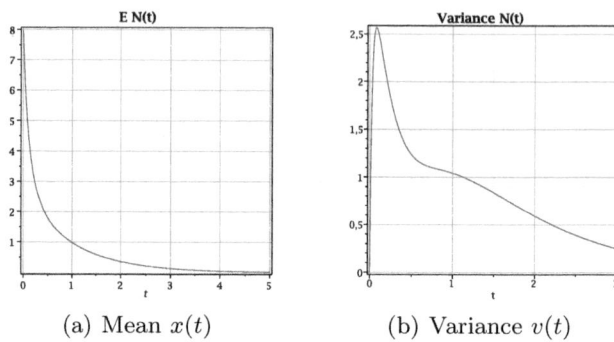

(a) Mean $x(t)$ (b) Variance $v(t)$

Fig. 8. Simulation of System (19), starting with state $n_0 = 8$

9   Conclusion

The algorithm presented in this paper allows the investigation on the study of genetic regulatory networks, considered from a stochastic point of view. The algorithm yields a system of differential equations whose integration yields values for some moments of the number of molecules of chemical species. The algorithm and its theory are formulated in the Weyl algebra. However, Proposition 2 shows how to replace computations in Weyl algebra by basic operations on commutative polynomials. This seems to produce a more efficient algorithm which combines the differentiation and evaluation steps of the straightforward approach. The issue of the infinite cascade, well-known is statistical physics, reduces the usefulness of the overall method. Approximation techniques, useful to break it, still need some investigation.

References

1. Abramov, S.A., Le, H.Q., Li, Z.: OreTools: a computer algebra library for univariate ore polynomial rings. School of Computer Science CS-2003-12, University of Waterloo (2003)
2. Chyzak, F.: The Ore_algebra library. In: Maple, Maplesoft, Canada. Software
3. Dixmier, J.: Enveloping Algebras. American Mathematical Society (1996); (Translation of the french edition Algèbres enveloppantes published in 1974 by Bordas)
4. Érdi, P., Tóth, J.: Mathematical models of chemical reactions. Princeton University Press (1989)
5. Feller, W.: An introduction to probability theory and its applications, 2nd edn., vol. I. John Wiley and Sons, Inc., New York (1957)
6. Gillespie, C.S.: Moment-closure approximations for mass-action models. Systems Biology, IET 3(1), 52–58 (2009)
7. Gillespie, D.T.: Exact Stochastic Simulation of Coupled Chemical Reactions. Journal of Physical Chemistry 81(25), 2340–2361 (1977)
8. Kalinkin, A.V.: Markov branching processes with interaction. Russian Math. Surveys 57, 241–304 (2002)

9. Klamt, S., Stelling, J.: Stoichiometric and Constraint-based Modeling. In: Szallasi, Z., Stelling, J., Periwal, V. (eds.) System Modeling in Cellular Biology: From Concepts to Nuts and Bolts, pp. 73–96. The MIT Press, Cambridge (2006)
10. Leykin, A.: D-modules for macaulay 2. mathematical software. In: Mathematical Software, pp. 169–179. World Sci. Publishing, River Edge (2002)
11. Ozbudak, M., Thattai, M., Kurtser, I., Grossman, A.D.: Regulation of noise in the expression of a single gene. Nature Genetics 31, 69–73 (2002)
12. Paulsson, J.: Models of stochastic gene expression. Physics of Live Rev. 2, 157–175 (2005)
13. Paulsson, J., Elf, J.: Stochastic Modeling of Intracellular Kinetics. In: Szallasi, Z., Stelling, J., Periwal, V. (eds.) System Modeling in Cellular Biology: From Concepts to Nuts and Bolts, pp. 149–175. The MIT Press, Cambridge (2006)
14. Reutenauer, C.: Aspects mathématiques des réseaux de Petri. Masson (1989)
15. Vidal, S.A.: Groupe Modulaire et Cartes Combinatoires. Génération et Comptage. PhD thesis, Université Lille I, France (July 2010)
16. Singh, A., Hespanha, J.P.: Lognormal moment closures for biochemical reactions. In: Proceedings of the 45th IEEE Conference on Decision and Control, pp. 2063–2068 (2006)
17. Tadao, M.: Petri nets: properties, analysis and applications. Proceedings of the IEEE 77(4), 541–580 (1989)

Reconciling Competing Models: A Case Study of Wine Fermentation Kinetics

Rodrigo Assar[1,*], Felipe A. Vargas[2], and David J. Sherman[1]

[1] INRIA Team MAGNOME and CNRS UMR 5800 LaBRI. Universit Bordeaux I
33405 Talence Cedex, France
[2] Department of Chemical and Bioprocess Engineering. Pontificia Universidad
Católica de Chile. Casilla 306, Correo 22, Santiago, Chile
{rodrigo.assar,david.sherman}@inria.fr, ftvargas@ing.puc.cl

Abstract. Mathematical models of wine fermentation kinetics promise early diagnosis of stuck or sluggish winemaking processes as well as better matching of industrial yeast strains to specific vineyards. The economic impact of these challenges is significant: worldwide losses from stuck or sluggish fermentations are estimated at 7 billion € annually, and yeast starter production is a highly competitive market estimated at 40 million € annually. Additionally, mathematical models are an important tool for studying the biology of wine yeast fermentation through functional genomics, and contribute to our understanding of the link between genotype and phenotype for these important cell factories.

We have developed an accurate combined model that best matches experimental observations over a wide range of initial conditions. This model is based on mathematical analysis of three competing ODE models for wine fermentation kinetics and statistical comparison of their predictions with a large set of experimental data. By classifying initial conditions into qualitative intervals and by systematically evaluating the competing models, we provide insight into the strengths and weaknesses of the existing models, and identify the key elements of their symbolic representation that most influence the accuracy of their predictions. In particular, we can make a distinction between main effects and secondary quadratic effects, that model interactions between cellular processes. We generalize our methodology to the common case where one wishes to combine competing models and refine them to better agree with experimental data. The first step is symbolic, and rewrites each model into a polynomial form in which main and secondary effects are conveniently expressed. The second step is statistical, classifying the match of each model's predictions with experimental data, and identifying the key terms in its equations. Finally, we use a combination of those terms to instantiate the combined model expressed in polynomial form. We show that this procedure is feasible for the case of wine fermentation kinetics, allowing predictions which closely match experimental observations in normal and problematic fermentation.

* Work supported by the doctoral program in Informatics of the University Bordeaux I, the INRIA Team MAGNOME and CNRS UMR 5800 LaBRI.

Keywords: wine fermentation, fermentation problems, combined and refined model, statistical comparison with experimental data, mechanistic kinetic models.

1   Introduction

Obtaining ways to combine models has become a necessity. When a development area has promoted interest of many investigation teams, it is common that competing models are developed, each one constructed to meet particular needs and that fit specific data. The same problem is approached of different ways, doing different estimations or working in different complexity levels. Reconciling models into a combined one it is possible to answer particular needs and obtaining general models. Different models may use different mathematical approaches and description formalisms, so we must conserve a degree of independence between the models.

In this particular study, we are interesting in the wine industry that annually produces 5 millions of tonnes only in France and whose worldwide losses from stuck or sluggish fermentations are estimated at 7 billion € annually. Yeast starter production is highly competitive (estimated at 40 million € annually). While fermentation has been used from antiquity for the production of wine and other alcoholic drinks, advances in the understanding of this process have allowed the modernization of this industry throughout time. As mechanisms that participate in fermentation are understood better it is possible to control the process. Fermentation is carried out by the action of yeasts, that convert the two grape sugars, glucose and fructose, to ethanol and by-products like aromas, flavors, carbon dioxide gas, and heat. Fermentation is the anaerobic alternative to respiration for generating energy molecules (ATP), but it is remarkably less efficient. While respiration generates ATP with yield of 30 per glucose molecule, fermentation produces only 2 ATP molecules. In spite of the efficiency of respiration, some types of yeasts, in particular *Saccharomyces cerevisiae*, prefer to ferment in the presence of oxygen given a high enough concentration of sugar. Because it is more rapid to get ATP from glycolysis and fermentation than from respiration, *S. cerevisiae* has developed biological mechanisms that, for high sugar concentrations, repress the synthesis of mRNA from genes that are involved in respiration and oxidizing metabolism, privileging fermentation ([5]). There exist several factors that affect the process of fermentation: the composition of the flora that is used to ferment, and those related with yeast nutrition and maintenance of viability of cells are fundamental. Although yeasts carry out the fermentation, some types of bacteria like *lactic, acetic bacteria* and *streptomyces* can survive fermentation ([10], [9]), affecting the results. Different strains and species of yeasts also influence the development of the process. Among the nutrients that are important are carbon sources (glucose, fructose) and nitrogen sources (amino acids, ammonia, nucleotide base peptides); temperature, presence of oxygen, low ethanol levels, and controlled pH are also important for the viability of cells. The pH index affect flora composition: low levels (< 3.5) inhibit many bacteria but yeasts are resistant to wide pH ranges. On the other hand, yeasts are tolerant to ethanol only in limited ranges, up to $17\%v/v$ approximately. Low temperatures favor non-yeast flora while high temperatures work against the diversity. Carbon and nitrogen sources are essential for the growth of yeast cells but their excess can generate problems too. One considers two types of fermentation problems in industrial applications: stuck and sluggish fermentation ([3]). Stuck fermentation happens when not all the sugar is consumed by the yeasts, and sluggish fermentations are those where the process is very slow. The risk of stuck fermentation is increased when the levels of temperature and sugar concentrations are high. Lower temperatures promote slower fermentation rates and consequently the risk of sluggish fermentation. High levels of nitrogen diminish the risk of sluggish fermentation but affects the flavor and quality of the wine. Strains of yeast that are highly tolerant to ethanol also have reduced risk of sluggish fermentation.

Different approaches have been used to explain the influence of fermentation factors on the production of ethanol. Even though the qualitative effect of the temperature or the inhibiting effect of the ethanol is known, it is not easy to quantify these relations in a mathematical model. A difficulty is the complexity of the biological mechanisms and the great amount of factors, making it necessary to estimate relations and reducing the number of variables. The variables that have been considered most important in previous studies are temperature, sugar, carbon dioxide, nitrogen, biomass, glycerol and ethanol. The influence of the type of yeast and mixtures have been studied little ([31], [23]). There exist two groups of models for the wine fermentation process; the first type focuses on the predictability and the second type on interpretability. By means of data mining techniques like decision trees, machine learning, support vector machines or neural networks ([29], [30], [33] and [32]), the first type of models exploit advances in computing technologies and large databases to predict fermentation profiles. These types of models have the advantage of including a large quantity of factors of the process, but they lack biological, physical or mechanical foundations and generally they are complex and difficult to interpret. The more predictable a the model, the more it is complex. The second type of models correspond to mechanistic kinetic models that are based on physical and biological principles. One of the first mechanistic models of oenological fermentation was developed by Boulton in 1980 ([4]), with the possibility of obtaining accurate measures of fermentation factors concentrations once the parameters of the models have been adjusted. This last approach allows the inclusion of black box models to estimate some behaviors, which also facilitates the generalization of the models to extend them to different environmental conditions, strains or cultures.

Our goal is to build a general method to build a combined model that reconciles existing models. We select models validated by their authors and look for the conditions in that they fit better reality, without modifying the internal coefficients of the models. For doing this, first we homogenize the notation and evaluate the models in different conditions and then we select the model that best represents reality in function of these conditions. Finally we build a combined model whose terms and coefficients are obtained of the original models. We applied our method to three interpretable models of the fermentation process: Coleman ([6]), Scaglia ([26]) and Pizarro ([22], [24]).

2   Methods

We divided our work in three steps: the symbolic, the statistical and the constructive one. First we rewrite each model to homogenize the notations and to separate the different effects that are included in the models. In particular for ODE models, we rewrite the models into polynomial form (or other appropriate base) to separate main and interaction effects. In the statistical step we classify each model according to how they agree with experimental data. For doing this, we identify the *factors* and select the *independent variables* that represent the process. To classify the results in function of configurations of factors, based in the availability of experimental data and the considerations of other studies, we construct a discreet set of intervals or levels of their domains. We compare the model's predictions of independent variables with experimental data for each configuration of factors. To decide the quality of the fitting we used two statistical criteria: confidence intervals and shape analysis. The first one is local and the second one is global. Given a configuration, one says that a simulation is *locally right* if it belongs to the confidence interval of experimental results for each time point (with experimental measures). This criterion allows us to decide if the model agrees with the experimental results at that time. For each time we can observe that there exist significant differences between simulations and experimental results, but the global behavior (the shape) can be the same. To decide if the shape of the simulated profile is similar to that of experimental results, we used linear regression and techniques of linearization to fit the experimental and simulated curves and compare them. We consider that a model is better than other one when the adjustment with experimental data is statistically better. Finally, we define a criterion to select the best model in function of configuration of factors and the independent variable to study. We build a combined model that optimizes the results to obtain the estimated profiles of the variables. This model uses coefficients of the original models and considers main and interaction effects.

Description of Analyzed Models

We analyzed three mechanistic approaches of fermentation process. The models have the common characteristic of being composed of a set of first degree differential equations, whose coefficients relate one variable with the others. Table 1 lists the number of equation associated to each variable and model in section of Results. The Coleman model ([6]) was built on the model previously presented by [7] to include the effects of temperature. Its main goal is introducing temperature dependency to predict difficult fermentations. It consists of a 5 coupled

Table 1. Number of equation associated to each variable and model

	Coleman ([6])	Pizarro ([22]).	Scaglia ([26])	Logistic ([20])
		Model		
X	1	6	11	15
X_A	2			
N	3	7		
$EtOH$	4	8	12	
S	5	9	13	
Gly		10		
CO_2			14	

ODEs (equations 1- 5) that are combined with 4 one-dimensional regression models to estimate parameters. The variables that are represented in differential equations are concentrations of: biomass (X), active biomass (X_A), nitrogen (N), ethanol ($EtOH$) and sugar (S). In basic terms, the Coleman model considers biomass concentration controlled by the growth rate and death rate (μ and τ). The growth rate is computed using Monod's equation with nitrogen nutrition and without considering competition ([19]; see Table 2); it considers that the lower is the remaining nitrogen the lower the growth rate. The death rate is considered proportional to ethanol concentration. The other fermentation variables are obtained by estimating production rates (for $EtOH$) or consumption rates (for N and S) per biomass unit. The effect of temperature is included to estimate parameters of the system with regression techniques. The Pizarro model ([22], [24], equations 6- 10) uses essentially the same differential equations as Coleman model, it adds the fermentation variable glycerol and does not consider active biomass concentration. The Pizarro method uses a different way for estimating uptakes and consumption rates per biomass unit. In this case intracellular behavior, studied by flux balance analysis, gives the specific production and consumption rates for the environmental conditions that are modeled by dynamic mass balance. It is built through an iterative process where intracellular network fluxes are bounded according to extracellular conditions, and for each iteration a maximization (of growth or glucose consumption rate) is performed to obtain uptakes and consumption rates that are used to predict extracellular concentrations of metabolites. The main goal is to better introduce the influence of environmental conditions in wine fermentation.

The Scaglia model includes other types of relations between variables (equations 11- 13). It considers only 4 fermentation variables: X (biomass concentration), S (sugar concentration), CO_2 (carbonic dioxide gas concentration) and $EtOH$ (ethanol concentration). The cell growth expression (equation 11) is based on Verlhust's logistic equation (15), where μ represents the growth rate and the quadratic coefficient of population, β, models the competition for available resources. The growth rate (μ in Table 2 for Scaglia method) of equation 15 is obtained by adaptations of Monod's model ([19]) with sugar nutrition: the lower is the remaining sugar the lower the growth rate. To model the death rate τ they observe that the faster the decrease of substrate concentration, the larger the increase in the cellular death rate. Scaglia included coefficients associated to proportion of yeast cells in growth cellular step, F_μ, and for those in death step, F_τ (see Table 2), to avoid the discontinuity of Blackman's equation to model growth rate ([28]). These factors estimate the transition according to the proximity of carbon dioxide emission to the maximum expected for a normal fermentation progress ($max(CO_2)$). It is considered Carbon dioxide concentration (CO_2, equation 14), being estimated with a emission rate coefficient per biomass unity, and the rate of an additional coefficient that we called $CO2Form$ (see Table 2). The ethanol production rate is obtained by estimating the conversion factor (yield) of carbon dioxide emitted to ethanol produced (equation 12). Nitrogen consumption is not considered, and the sugar consumption rate (equation 13) is composed by a term acting on biomass and a quadratic term (logistic equation type, 15).

The three teams worked on different strains of *Saccharomyces cerevisiae*. The Coleman team worked with the yeast *Premier Cuvee* (Red Star, Milwaukee, WI). It is a commercial strain of *Saccharomyces cerevisiae*. The fermentations were prepared at pH 3.35, total nitrogen was determined by measuring ammonia concentration and alpha amino acid concentration. The Pizarro team worked with *Prise de Mousse* EC1118 (Lalvin, Zug, Switzerland), which is another commercial strain. The pH was kept constant at 3.5, nitrogen measures considered ammonia and free aminoacidic nitrogen. The Scaglia team used experimental results of other two studies: [31] and [11]. The *Saccharomyces cerevisiae Bsc411* of [31] was identified according to *Kurtzman & Fell* protocols and was taken from Argentina. The Fleet team studies are widely recognized. In both cultures acidity was controlled, the reducing sugars were determined colorimetrically using *the 3,5-dinitrosalicylic acid (DNS) method* ([18]). Coleman and Pizarro models reviewed an ample range of temperatures and initial conditions of sugar and nitrogen. The Scaglia model was adjusted only on moderate levels of temperature, sugar and nitrogen concentrations. For more details about cultures and fermentation conditions which were used, refer directly to the papers.

Experimental Data

In our study we considered experimental data of three papers: [22], [15] and [16]. The experimental measures of the Pizarro team correspond to a wide range of data. We used laboratory results that were obtained with the strain *Prise de Mousse EC1118*, and industrial results for Industrial Cabernet Sauvignon, wine fermentations that were monitored during the 2003 vintage at a commercial winery in Chile. Sugar profiles for six batch fermentations at 28 °C with high/low nitrogen and other two at 12 °C and 17 °C with high conditions of nitrogen were used to calibrate the model, and consequently we expect a better adjustment of this model for sugar in this conditions. By direct communication with [15], we obtained data of biomass profiles in two particular conditions: moderate temperature (24 °C), high level of sugar (280 g/l) and moderate/high levels of nitrogen (approximately 220 mg/l and 551 mg/l respectively). The data set available in Mendes-Ferreira studie ([16]) used the strain *Saccharomyces cerevisiae*

PYCC4072 that was supplied by the Portuguese Yeast Culture Collection. The paper describes experimental biomass, ethanol and sugar (and other indexes) results for two experimental conditions, fermentation maintained at 20 °C with moderate initial sugar concentration (200 g/l) and initial nitrogen concentration high (267 mg/l) or low (66 mg/l). The acidity conditions were adjusted to pH 3.7, nitrogen is supplied by ammonium phosphate and sugar corresponds to glucose.

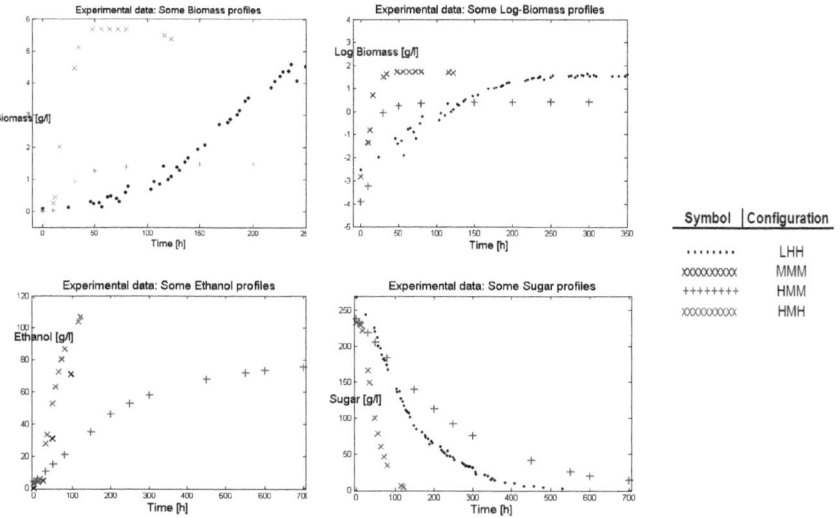

Fig. 1. Fermentation variables profiles for some initial conditions configurations. For Biomass and Sugar one shows experimental results in *LHH* (low temperature: 12 °C, high initial sugar: 268 g/l, high initial nitrogen: 300 mg/l), *HMM* (28°C, sugar: 238 g/l, nitrogen: 50mg/l) and *HMH* (28 °C, sugar: 233 g/l, nitrogen: 300 mg/l) configurations. For Ethanol we show data for *MMM* configuration (20 °C, sugar: 200 g/l, nitrogen: 66 mg/l) instead of *LHH*. Log-Biomass profiles are shown too, we obtained linear correlations in transient phase of 0.98, 0.99 and 0.97 respectively.

3   Results

Exploratory Analysis

The observation of experimental data gave us some ideas about the profiles of fermentation variables. In the three variables (biomass, ethanol and sugar concentrations) we observed two phases, transient and stable. Before a particular time, that we call *stabilization time*, fermentation variables change exponentially over time. After stabilization these are statistically constant. We verified the exponential behavior of biomass profiles statistically by means of linearization and linear regression (Figure 1), the growth rate can be assumed constant over time. In the case of biomass, in the first phase it increases exponentially until the cells stop

Table 2. Notation, comparison with original papers nomenclature

Notation	Meaning	Coleman ([6])	Pizarro ([22])	Original notation and computation Scaglia ([26])
X	viable biomass concentration, $[g \cdot l^{-1}]]$	X total biomass concentration	X_V	X
X_A	active biomass concentration, $[g \cdot l^{-1}]]$	X_A		
N	nitrogen concentration, $[g \cdot l^{-1}]$	N	NH_4	N
$EtOH$	ethanol concentration, $[g \cdot l^{-1}]$	E	$EtOH$	P
S	sugar concentration, $[g \cdot l^{-1}]$	S	glu	S
Gly	glycerol concentration, $[g \cdot l^{-1}]$	Gly		
CO_2	carbon dioxide concentration, $[g \cdot l^{-1}]$			CO_2 $12.072 \cdot \dfrac{max(\mu) \cdot S}{S + K_S \cdot 93.023^{1.508}}$
μ	growth rate of yeast cells, $[h^{-1}]$	Monod's model nitrogen nutrition: $\dfrac{max(\mu) \cdot N}{K_N + N}$	μ, Flux balance analysis	$(0.0001 - 0.047 \cdot \dfrac{dS}{dt})$
τ	death rate of yeast cells, $[h^{-1}]$	k_d, temperature and ethanol dependent		
ν_N	nitrogen consumption rate per yeast mass, $[g \cdot g^{-1} \cdot h^{-1}]$	$\dfrac{\mu}{Y_{X/N}}$	ν_{NH_4}, Flux balance analysis	
ν_{EtOH}	ethanol production rate per yeast mass, $[g \cdot g^{-1} \cdot h^{-1}]$	$\beta \dfrac{max(\nu_{EtOH}) \cdot S}{K_S + S}$	$= \nu_{EtOH}$, Flux balance analysis	
ν_S	sugar consumption rate per yeast mass, $[g \cdot g^{-1} \cdot h^{-1}]$	$\dfrac{\nu_{EtOH}}{Y_{EtOH/S}}$	ν_{glu}, Flux balance analysis	$Y_{X/S}^{-1} \cdot \dfrac{max(\mu) \cdot S}{S + K_S \cdot 93.023^{1.508}}$
ν_{S_0}	initial sugar consumption rate per yeast mass, $[g \cdot g^{-1} \cdot h^{-1}]$			$F = 0.008$, initial sugar dependent
ν_{gly}	glycerol production rate per yeast mass, $[g \cdot g^{-1} \cdot h^{-1}]$		ν_{gly}, Flux balance analysis	
ν_{CO2}	carbon dioxide production rate per yeast mass, $[g \cdot g^{-1} \cdot h^{-1}]$			$G \cdot \dfrac{max(\mu) \cdot S}{S + K_S \cdot 93.023^{1.445}}$
K_N K_S	constant for nitrogen-limited growth, $[g \cdot l^{-1}]$ constant for sugar utilization in growth, $[g \cdot l^{-1}]$	0.009 Monod constant 10.278 Michaelis-Menten-type constant		2.15 Monod constant for sugar-limited growth
$Y_{X/N}$	yield coefficient for cell mass grown per mass of nitrogen used, $[g \cdot g^{-1}]$	$Y_{X/N}$, temperature dependent		
$Y_{EtOH/S}$	yield coefficient for ethanol produced per sugar consumed, $[g \cdot g^{-1}]$	$Y_{E/S} = 0.55$		
$Y_{X/S}$	yield coefficient for cells formed per sugar consumed, $[g \cdot g^{-1}]$			0.029
$Y_{CO2/EtOH}$	yield coefficient for CO_2 formed per ethanol produced, $[g \cdot g^{-1}]$			$Y_{CO2/P}$, initial conditions dependent
$max(\mu)$	maximum growth rate of yeast cells, $[h^{-1}]$	μ_{max}, temperature dependent		μ_m, initial conditions dependent
$max(\nu_{EtOH})$	maximum ethanol production rate, $[g \cdot g^{-1} \cdot h^{-1}]$	β_{max}, temperature dependent		
$max(CO_2)$	maximum CO_2 for normal fermentation progress, $[g \cdot l^{-1}]$			$CO_{2,95} = 78$, corresponds to 95% of maximum emissions for initial conditions.
F_μ	correction factor of growth rate			$\dfrac{exp(-(CO_2 - max(CO_2)))}{exp(-(CO_2 - max(CO_2))) + exp(-(CO_2 - max(CO_2)))}$
F_τ	correction factor of death rate			$\dfrac{1}{1 - F_g}$
β	β competition coefficient in logistic equation 15, $[l \cdot g^{-1}]$			β, initial conditions dependent
$CO2Form$	additional Scaglia concentration coefficient for carbon dioxide formation, $[g \cdot l^{-1}]$			$241.44 \cdot max(\mu) \dfrac{S^2}{(S + K_S \cdot 93.023^{1.0})(S + K_S \cdot 93.023^{1.445})}$ $X + 0.01 \cdot X$

their growth. Ethanol concentration increases while the yeast cells are active, after the stabilization time the production stops. Sugar concentration decays in an exponential way until it is completely consumed. Different samples show different uptake (for biomass and ethanol) and consumption rates (for sugar). For the Coleman ([6]) and Pizarro ([22]) models, fermentation variables evolve in time according to uptake (biomass, ethanol and glycerol) or consumption (sugar, nitrogen) factors per concentration unity of yeast cell (biomass). They assume that these coefficients change in time depending of the fermentation or environmental variables and do not depend only on initial conditions. When solving these models one obtains exponential behaviors whose rates change over time, according to the value of the fermentation variables or environmental conditions, finishing in a stable phase. According to the sign of the factors we have exponential growth (positive sign) or decay (negative sign) followed by stabilization. Biomass profiles resemble the solutions of logistic differential equation 15, that are well known in ecology to model population growth. These types of differential equations were derived by Verhulst in 1838 to describe the self-limiting growth of a biological population (Verlhurst's model; see [20]). Population starts to grow in an exponential phase, as it gets closer to the carrying capacity the growth slows down and reaches a stable level. The equations (11-13) that define the Scaglia model ([26]) include logistic components but they are more complex, and one observes relations between one-order differential expressions of variables.

Symbolic Step: Rewriting the Models

We rewrote the models into a polynomial form that in this case allowed to separate main effects of interaction effects that are represented by quadratic coefficients. The Coleman model is expressed by equations 1-5, the Pizarro model by equations 6-10 and the Scaglia model by equations 11-14. Table 2 describes the meaning of the variables and our notation as compared with the nomenclature of the original papers.

$$\frac{dX}{dt} = \mu \cdot X_A \tag{1}$$

$$\frac{dX_A}{dt} = (\mu - \tau) \cdot X_A \tag{2}$$

$$\frac{dN}{dt} = -\nu_N \cdot X_A \tag{3}$$

$$\frac{d[EtOH]}{dt} = \nu_{EtOH} \cdot X_A \tag{4}$$

$$\frac{dS}{dt} = -\nu_S \cdot X_A \tag{5}$$

$$\frac{dX}{dt} = \mu \cdot X \tag{6}$$

$$\frac{dN}{dt} = -\nu_N \cdot X \tag{7}$$

$$\frac{d[EtOH]}{dt} = \nu_{EtOH} \cdot X \tag{8}$$

$$\frac{dS}{dt} = -\nu_S \cdot X \tag{9}$$

$$\frac{d[Gly]}{dt} = \nu_{Gly} \cdot X \tag{10}$$

$$\frac{dX}{dt} = (F_\mu \cdot \mu - F_\tau \cdot \tau) \cdot X - F_\mu \cdot \beta \cdot X^2 \tag{11}$$

$$\frac{d[EtOH]}{dt} = \frac{1}{Y_{CO_2/EtOH}} \cdot \frac{dCO_2}{dt} \tag{12}$$

$$\frac{dS}{dt} = -\left((\nu_S + \nu_{S_0}) \cdot X - \frac{0.00002}{Y_{X/S}} \cdot X^2\right) \tag{13}$$

$$\frac{dCO_2}{dt} = \nu_{CO_2} \cdot X + \frac{d(CO2Form)}{dt} \tag{14}$$

$$\frac{dX}{dt} = \mu \cdot X \cdot \left(1 - \frac{\beta}{\mu} \cdot X\right) \tag{15}$$

Statistical Step: Classifying the Results

For fermentation process the considered factors are initial conditions and time. We studied the fermentation variables: X (biomass concentration), $EtOH$ (ethanol concentration) and S (sugar concentration). Initial conditions were determined by ranges of initial temperature, sugar and nitrogen concentration, and we divided in transient and stable phase. The levels of initial conditions were ranges of initial temperature, sugar and nitrogen concentrations. We considered that the temperature is *low* when it is lower than 19 $°C$, is *moderate* for values between 20 $°C$ and 27 $°C$, and *high* for larger values. Initial sugar concentration was called *moderate* for values less than 240 g/l and *high* for superior values. Initial nitrogen concentration was *moderate* for values less than 240 mg/l and *high* for those superior values. For each initial conditions configuration, we separated the profiles in *transient* and stable *phase* by analyzing the experimental results.

In Table 3 we show the origin of experimental data for different initial condition levels and fermentation variables. This classification allowed us to cover a wide range of configurations. In spite of this, for some combinations we do not have experimental data because conditions of fermentation are difficult. This is the case for *low* initial temperature, sugar and nitrogen concentrations. For *high* temperature and sugar with insufficient levels of nitrogen source, we have the same situation. Part of the data have superior statistical quality. While the number of samples that describe an initial condition configuration is larger, the quantity of information that validates our assertions about the adjustment of each model in this configuration is also larger. In particular the Pizarro sample for HMM (high temperature, moderate initial sugar level and moderate initial nitrogen level) and HMH configurations give us standard deviations for

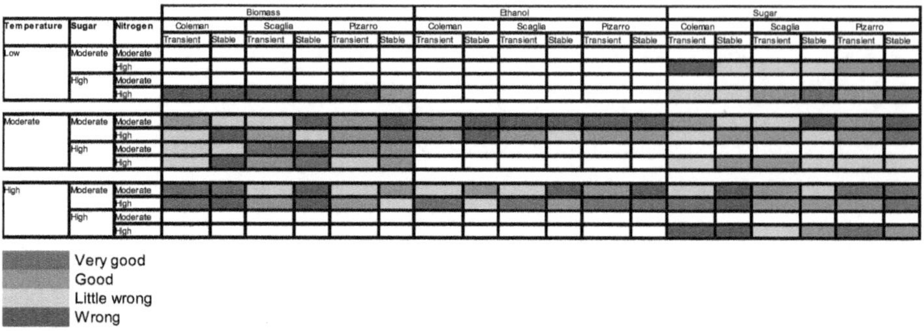

Fig. 2. Summary of results of adjustment according to initial conditions. Quality of adjustment of each model for each initial conditions configuration and phase (transient and stable).

variable profiles. Mendes-Ferreira ([16]) samples for MHM (moderate temperature, high initial sugar level and moderate initial nitrogen level) and MHH supply means and standard deviations of measures too. We evaluated the three

Table 3. Origin of experimental data. For each Initial Temperature-Sugar-Nitrogen configuration, and fermentation variable it is showed the origin of available data.

Temperature	Sugar	Nitrogen	Biomass	Ethanol	Sugar
Low (0-19 °C)	Moderate (160-240 [g/l])	Moderate (50-240 [mg/l])			1
		High (240-551 [mg/l])			
	High (240-308 [g/l])	Moderate			
		High	2		3
Moderate (20-27 °C)	Moderate	Moderate	1	1	1
		High	1	1	6 1
	High	Moderate	1		
		High	1		2
High (28-35 °C)	Moderate	Moderate	1		1
		High	2	1	2
	High	Moderate			
		High			1

Pizarro data ([22])
Malherbe data ([15])
Mendes-Ferreira data ([16])

studied models according to how well they agree with the experimental results (Figure 2). For each sample we reviewed the adjustment in the transient and the stable phase. For the local criterion, at each point we built confidence intervals of experimental results by using measures of means and standard deviations, and computed the *p-values* associated to the decision of considering simulated value equal to experimental result. Because we observe exponential behavior, for the global criterion we have computed the correlation between the logarithm of simulations and the data over the time (Table 4). In general, an adjustment was considered *Very good* if the local criterion and the global one are very favorable ($p\text{-}value \geq 0.1$, $correlation \geq 0.98$); *Good* if a criterion is very favorable and the other one is only favorable ($0.05 \leq p\text{-}value < 0.1$ or $0.95 \leq correlation < 0.98$);

Little wrong if a criterion is unfavorable (*p-value* < 0.05 or *correlation* < 0.95) and the other one is favorable or superior; and *Wrong* if both criteria are unfavorable. The cases near to the limits were checked especially. In case the local criterion is absolutely unfavorable (*p-value* = 0) we qualified in *Wrong*, if local criterion is unfavorable (but not absolutely) and global criterion is optimum (*correlation* = 1) we considered it *Good*.

Table 4. Statistical analysis of models. For each configuration of factors and fermentation variable, we show the average *p-value* for the local criterion (C.1), and *correlation* for global criterion (C.2). The lower the *p-value*, the bigger the local error in simulations. The bigger the *correlation*, the bigger the global similarity between data and simulations. Results for other configurations can be asked.

		Coleman model				Pizarro model				Scaglia model				Combined model			
		Transient		Stable		Transient		Stable		Transient		Stable		Transient		Stable	
Config.	Variable	C.1	C.2	C.1	C.2	C.1	C.2	C.1	C.2	C.1	C.2	C.1	C.2	C.1	C.2	C.1	C.2
MMM	X	0.006	1	0.006	0.92	0.109	1	0	1	0.012	0.96	0	0.77	0.033	1	0.006	0.92
	$EtOH$	0.001	1	0	0.97	0	1	0	1	0	1	0	0.93	0.001	1	0	0.97
	S	0.07	0.97	0.001	0.93	0.022	1	0	0.98	0.017	0.95	0	0.98	0.063	1	0.001	0.93
MMH	X	0.045	0.98	0.001	0.90	0.006	1	0.483	1	0.113	0.98	0.06	0.99	0.256	0.99	0.483	1
	$EtOH$	0.047	1	0	0.97	0.001	1	0.053	1	0.048	1	0.149	0.92	0.049	1	0.053	1
	S	0.129	0.93	0.152	0.98	0.140	0.98	0.27	0.98	0.064	0.99	0.015	0.97	0.129	0.99	0.27	0.98
HMM	X	0.388	0.99	0.476	0.97	0.102	0.83	0.027	1	0.082	0.99	0	0.87	0.388	0.99	0.476	0.97
	$EtOH$	0.049	0.95	0.155	0.97	0.129	0.95	0	1	0.08	0.92	0	-0.97	0.129	0.95	0.155	0.97
	S	0.171	0.79	0	-0.7	0.238	1	0.272	0.97	0.107	0.91	0.032	0.73	0.238	1	0.272	0.97
HMH	X	0.203	0.97	0	0.28	0.156	0.98	0.048	0.97	0.197	0.96	0	0.91	0.203	0.97	0.048	0.97
	$EtOH$	0.275	1	0.089	0.80	0.162	0.99	0.214	0.99	0.264	0.98	0.001	1	0.339	0.99	0.214	0.99
	S	0.327	1	0	0.59	0.135	0.98	0.167	0.99	0.197	0.95	0.001	1	0.327	1	0.167	0.99

Constructive Step: Building the Combined Model

As a direct consequence this criterion allows us to obtain better predictions of fermentation variable profiles than those obtained by each individual model. In Table 5 we summarize our criteria, and according to initial conditions and variable to consider (between Biomass, Ethanol and Sugar) we say which approach to use. We used the criteria to build a combined model where we separate the different effects in polynomial way to obtain a *combined model* (equations 16-18). These equations capture the three models and the variables X_A and CO_2 are those computed by the Coleman and Scaglia models respectively. The coefficients μ_A, ϵ_A and σ_A correspond to coefficients of the Coleman model to represent linear effect of X_A on X. The linear coefficient of X on $EtOH$, $\epsilon^{(1)}$, is associated to the Pizarro model; $\mu^{(1)}$ is composed by contributions of the Pizarro and Scaglia models, and quadratic effects (coefficients $\mu^{(2)}$ and $\sigma^{(1)}$) are obtained from the Scaglia model. The coefficients are active or not in function of initial configuration and time (Table 6). For instance let us consider the configuration MMH. The equations 16-18 take the form from the equations 19-21. One observes that the main factor is the linear effect of X. In the transient phase there exists a cuadratic effect of X, a linear effect of CO_2 rate on $EtOH$, and X_A only affects the ethanol modelling.

Table 5. Criterion of selection of best models in function of initial conditions. For each combination variable-phase is written the best model, colors represent the quality of the adjustment when comparing between all the initial conditions.

Temperature	Sugar	Nitrogen	Biomass Transient	Biomass Stable	Ethanol Transient	Ethanol Stable	Sugar Transient	Sugar Stable
Low (0-19 °C)	Moderate (180-240 [g/l])	Moderate (50-240 [mg/l])						
		High (240-551 [mg/l])					Pizarro	Pizarro
	High (240-308 [g/l])	Moderate						
		High	Indiferent	Pizarro			Scaglia/Pizarro	Scaglia/Pizarro
Moderate (20-27 °C)	Moderate	Moderate	Coleman/Pizarro	Coleman	Coleman	Indiferent	Coleman/Pizarro	Coleman
		High	Scaglia/Pizarro	Pizarro	Indiferent	Pizarro	Scaglia/Pizarro	Pizarro
	High	Moderate	Scaglia	Pizarro			Scaglia	
		High	Scaglia	Scaglia			Scaglia	Coleman
High (28-35 °C)	Moderate	Moderate	Coleman	Coleman	Pizarro	Coleman	Pizarro	Pizarro
		High	Coleman	Pizarro	Indiferent	Pizarro	Coleman	Pizarro
	High	Moderate						
		High					Pizarro	Scaglia/Pizarro

Quality	Worst							Best

Table 6. Temporal intervals at which the coefficients of equations 16-18 are active for each configuration of initial conditions and the formulas. One writes − if there are not experimental data to validate, FBA denotes the result by using Flux balance Analysis.

Coefficient	Meaning	LMH	LHH	MMM	MMH	MHM	MHH	HMM	HMH	HHH
$\mu_A = \frac{max(\mu)\cdot N}{K_N+N}$	linear effect of X_A	-	∅	∀t	∅	∅	∅	∀t	$t \le 30$	-
$\mu^{(1)} = FBA,\ (F_\mu \cdot \mu + F_T \cdot \tau)$	linear effect of X	-	$t > 110$	$t \le 96$	∀t	∀t	∀t	∅	$t > 30$	-
$\mu^{(2)} = F_\mu \cdot \beta$	quadratic effect of X	-	∅	∅	$t \le 51$	$t \le 27$	∀t	∅	∅	-
$\epsilon_A = \frac{max(\nu_{EtOH})\cdot S}{K_S+S}$	linear effect of X_A	-	-	$t \le 96$	$t \le 51$	-	-	$t > 300$	$t \le 30$	-
$\epsilon^{(1)} = FBA$	linear effect of X	-	-	∅	∀t	-	-	$t \le 300$	∀t	-
$\epsilon_{CO_2} = \frac{1}{Y_{CO_2/EtOH}}$	linear effect of $\frac{dCO_2}{dt}$	-	-	∅	$t \le 51$	-	-	∅	$t \le 30$	-
$\sigma_A = \frac{\nu_{EtOH}}{Y_{EtOH/S}}$	linear effect of X_A	∅	∅	∀t	∅	-	$t > 107$	∅	$t \le 30$	∅
$\sigma^{(1)} = FBA,\ 0.008 + \frac{max(\mu)\cdot S}{Y_{X/S}\cdot(S+K_S\cdot 93.02^{1.51})}$	linear effect of X	∀t	∀t	$t \le 96$	∀t	-	$t \le 107$	∀t	$t > 30$	∀t
$\sigma^{(2)} = \frac{2\cdot 10^{-5}}{Y_{X/S}}$	quadratic effect of X	∅	∀t	∅	$t \le 51$	-	$t \le 107$	∅	∅	$t > 103$

$$\frac{dX}{dt} = \mu_A \cdot X_A + \mu^{(1)} \cdot X - \mu^{(2)} \cdot X^2 \tag{16}$$

$$\frac{d[EtOH]}{dt} = \epsilon_A \cdot X_A + \epsilon^{(1)} \cdot X + \epsilon_{CO_2} \cdot \frac{dCO_2}{dt} \tag{17}$$

$$\frac{dS}{dt} = -\left(\sigma_A \cdot X_A + \sigma^{(1)} \cdot X - \sigma^{(2)} \cdot X^2\right) \tag{18}$$

$$\frac{dX}{dt} = \mu^{(1)} \cdot X - 1_{t\le51} \cdot \mu^{(2)} \cdot X^2 \tag{19}$$

$$\frac{d[EtOH]}{dt} = 1_{t\le51} \cdot \epsilon_A \cdot X_A + \epsilon^{(1)} \cdot X + 1_{t\le51} \cdot \epsilon_{CO_2} \cdot \frac{dCO_2}{dt} \tag{20}$$

$$\frac{dS}{dt} = -\left(\sigma^{(1)} \cdot X - 1_{t\le51} \cdot \sigma^{(2)} \cdot X^2\right) \tag{21}$$

For configuration MMH we observed an initial effect of competition to consume resources; but for HHH the sugar consumption (equation 22) initially, when the substrates are abundant, competition does not exist but it appears when the resources become scarce.

$$\frac{dS}{dt} = -\left(\sigma^{(1)} \cdot X - 1_{t>103} \cdot \sigma^{(2)} \cdot X^2\right) \qquad (22)$$

As a result, in function of factors, we go from one profile type to another one. Temporal phase and initial condition affects the results. For each fermentation there exists a time at which the profiles change from transient to stable phase: *stabilization time*. According to environmental conditions one obtains different profiles of the fermentation variables; for different level of initial temperatures, sugar and nitrogen concentration, one observes different growth/decrease rates and the change to the stable phase happens in different time (Figure 1). The bigger the initial temperature, sugar or nitrogen concentration; the bigger the growth rate of biomass. As the biomass changes to its stable phase, ethanol production and the consumption of sugar stop.

4   Conclusions and Discussion

We built a general method to combine models in function of configurations of factors. The method was applied to fermentation process modelling to explain the profiles of fermentation variables: concentration of yeast biomass, ethanol and sugar; by considering four factors: initial temperature, sugar and nitrogen, and growth phase. Our method starts with a symbolic step to homogenize the notation, for ODE models by rewriting into polynomial form and by identifying main and interaction effects. It continues with a statistical step to evaluate the models, in function of experimental data ([22], [15] and [16]). We defined discrete levels for each factor, for each configuration of factors and fermentation variable we statistically compared the results of three kinetic fermentation models ([6], [22] and [26]) with the experimental results and we obtained quality indexes of each model (Figure 2). We finished with the construction of a *combined model*, where one selects the best resolution method for each fermentation variable and configuration of factors (Table 5). The equations 16-18 allows to interpret the combined model in function of initial configuration, for instance equations 19-21 for MHH. Although generally for all variables there exist combinations of models of good adjustment to the experimental data, for each one of the fermentation variables and initial conditions the approaches showed different quality levels (Table 5, Figure 2). The best simulations were obtained for sugar consumption, in general terms Pizarro model showed the best adjustments especially for *high* levels of temperature. For *low* temperature and *high* levels of sugar and nitrogen the Scaglia model showed results similar to those of the Pizarro model. The configuration MMH (Figure 3) was covered by two different data sets for sugar simulations, the Pizarro and Mendes-Ferreira data ([16]). We observed similar measures between data sets and that for the transient phase

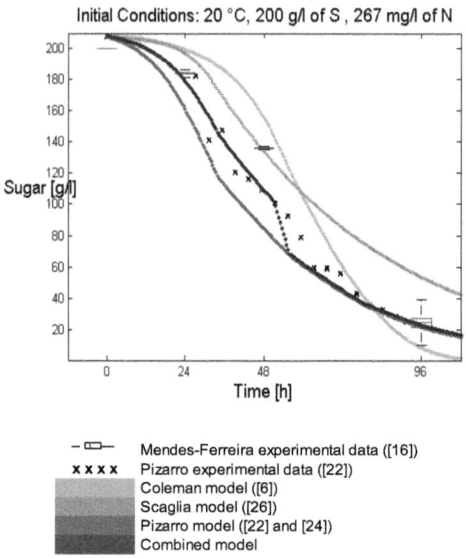

Fig. 3. Comparison between models adjustments for sugar profiles in MMH initial configuration. Sugar profiles simulated by the three models and our combined model. Two experimental samples; with temperature: 20 $°C$, sugar: 200 and 207 g/l, nitrogen: 267 and 240 mg/l.

the Scaglia model agreed better with the Mendes-Ferreira data ([16]), and for the stable phase the predictions of Pizarro are the best. For the HMM configuration we observed the best results with the Pizarro simulations; this agreed with the calibration data used to estimate sugar uptake parameters. For HMH configuration Coleman model showed the best results in transient phase, Pizarro worked better in stable phase and for HHH configuration.

The worst quality levels were obtained for biomass: Coleman and Scaglia models best agreed with experimental data in the transient phase. The configurations HMH and HMM showed the best results for Coleman model; MHM and MHH for the Scaglia model. Pizarro model worked better in stable phase, best results in LHH, MMH, MHM, MHH and HMM. Pizarro and Coleman models showed the best results for Ethanol production simulations.

Our combined model, obtained good results for almost all the initial conditions configurations. As it is observed in Figure 2 there exist very few initial conditions in which no model obtains good results of adjustment. The only negative cases are LHH configuration in transient phase for biomass, and MMM configuration in stable phase of Ethanol. We observed the best results for sugar profiles simulations, for all the initial conditions one obtains that the transient or stable phase is adjusted with quality *good* (Table 5). For the transient phase the best configurations are HHH, HMM and HMH; for stable phase HMM, followed by LMH and LHH. For this fermentation variable, the configurations LHH, HMM and HMH were represented mainly by Pizarro model, in which they were used to calibrate sugar-uptake parameters. For MMH (Figure 3) we obtained *good* results in transient phase, by combining Scaglia and Pizarro model, and *very good* results in stable phase with the Pizarro model. The best result for biomass adjustments were obtained for the configuration HMM that showed very good results for both transient and stable phase. We obtained good results for MMH, MHM and MHH in both temporal phases too. With respect to simulations of ethanol profiles, the best result was obtained for HMH configurations. For this configuration the transient phase is very good represented by the three models and the stable phase only by the Pizarro approach. Another important fermentation variable is nitrogen concentration. The experimental data of [15] and [16] give us nitrogen measurements for MMM, MMH, MHM and MHH configurations. We observed that for these initial conditions, the combined model is completely represented by the Pizarro model. It obtains the best results (good or very good quality), the Coleman approach showed poor results and the Scaglia model does not include this variable.

In this study we chose to build combined models without changing the individual models, since by design we assume that the original models are validated. The other alternative is to tune the internal parameters to refine the models and to obtain more generality. Several types of experimental results can be included to improve the estimation methods, looking for the correct inclusion of the relevant factors of the fermentation process. The effect of these factors can directed computed by using devices to measure the number of cells (by using Hemacytometer or Neubauer for example) on fermentation samples or by biotechnological tools as *DNA Microarrays* ([14]) and *PCR* (Polymerase chain reaction, [13]). The estimations of profiles features as growth rate, death rate and yield coefficients developed in [6] can be extended to more strains and species. Microarrays and PCR can give us estimations of the profiles too. In [1], [16] and [17] were done studies of gene expression profiles of particular strains of *Saccharomyces cerevisiae* during fermentations with high level of sugar but different levels of nitrogen concentration. In [1] was observed that some genes involved in biosynthesis of macromolecular precursors have superior levels of expression in high nitrogen condition than low. Low levels of nitrogen showed expression levels superior for genes involved in translation and oxidative carbon metabolism. In [16] and [17] were observed early responses of yeast cells to low nitrogen. They identified 36 genes highly expressed under conditions of low or absent nitrogen in comparison with a nitrogen-replete condition for *Saccharomyces cerevisiae PYCC4072*, the behavior of four of these transcripts was confirmed by *RT-PCR* analysis in this and another wine yeast strain. The signature genes of both studies can be used to predict nitrogen deficiency and to prevent fermentation problems. These ideas can be extended to study the temporal transcriptional, responses of genes on different pH, initial temperature, sugar and nitrogen concentrations, strains and species of yeasts. The analysis can be oriented, for example, to genes associated with oxidation of glucose, glycolysis and anaerobic functions. Expression levels of enzymes allow to simulate phenotype by FBA on metabolic models to obtain uptakes and consumption rates on different conditions and yeast strains.

Another challenge to obtain better approaches of the reality is to construct and to calibrate fermentation models that consider interacting yeast populations competing by resources. Although it has been observed that *Saccharomyces cerevisiae* is dominant in the majority of spontaneous alcoholic fermentations ([12], [25]) and it is the most popular yeast in inoculated cultures, there exist many strains and other yeasts as *Candida cantarellii* that participate in the process ([21], [34]) and it influences the aroma ([23]). The intervention of *lactic acid and acetic bacteria* in fermentations is also documented ([10], [9]). One can consider competence between individuals of the same population, modelled by logistic-like models similar to equation 15, and interactions between different populations. A usual way to model the presence of two or more populations competing by resources is the *Lotka-Volterra-like models* ([27]). Different strains of *Saccharomyces cerevisiae* can present different levels of tolerance to ethanol, acidity, growth and death rate between other coefficients, Another fermenting yeast *Candida cantarellii* present different rates of growth, ethanol and glycerol yields [31]. In future versions we will introduce the dependency of these rates with respect to yeast strains or species, and pH conditions.

In *System Biology* the combination of models takes relevance to analyze hierarchical systems. The notion *composition* ([8]) is used to build models by defining their components and the relations between them. A system is analyzed in a hierarchic way, defining it as being composed by subsystems. The components with different nature are well defined using different formalisms to generating sub-models, and *Base formalisms* capable of including the semantics of a wide variety of languages are used to define combined models ([2]). In this study we focused in the mathematical way of combining models, in future works we will approach the formalism for defining general combinations.

References

1. Backhus, L.E., DeRisi, J., Brown, P.O., Bisson, L.F.: Functional genomic analysis of a commercial wine strain of saccharomyces cerevisiae under diering nitrogen conditions. FEMS Yeast Research 1, 111–125 (2001)
2. Barros, F.J., Mendes, M.T., Zeigler, B.P.: Variable DEVS-variable structure modeling formalism: an adaptive computer architecture application. In: Fifth Annual Conference on AI, and Planning in High Autonomy Systems, Gainesville, FL, USA, pp. 185–191 (1994)
3. Bisson, L.F.: Stuck and sluggish fermentations. Am. J. Enol. Vitic. 50(1), 107–119 (1999)
4. Boulton, R.: The prediction of fermentation behavior by a kinetic model. Am. J. Enol. Vitic. 31(1), 40–45 (1980)
5. Boulton, R.B., Singleton, V.L., Bisson, L.F., Kunkee, R.E.: Principles and Practices of Winemaking, 1st edn. Springer, Heidelberg (1996)
6. Coleman, M.C., Fish, R., Block, D.E.: Temperature-Dependent kinetic model for Nitrogen-Limited wine fermentations. Applied and Environmental Microbiology 73(18), 5875–5884 (2007); PMID: 17616615 PMCID: 2074923
7. Cramer, A.C., Vlassides, S., Block, D.E.: Kinetic model for nitrogen-limited wine fermentations. Biotechnology and Bioengineering 77(1), 49–60 (2002)

8. SBML developers. Sbml composition workshop (September 2007), `http://sbml.info/Events/Other_Events/SBML_Composition_Workshop_2007`

9. Drysdale, G.S., Fleet, G.H.: Acetic acid bacteria in winemaking: A review. Am. J. Enol. Vitic. 39(2), 143–154 (1988)

10. Fleet, G.H., Lafon-Lafourcade, S., Ribreau-Gayon, P.: Evolution of yeasts and lactic acid bacteria during fermentation and storage of bordeaux wines. Applied and Environmental Microbiology 48(5), 1034–1038 (1984); PMID: 16346661 PMCID: 241671

11. Fleet, G.H.: Wine Microbiology and Biotechnology, 1st edn. CRC Press (1993)

12. Frezier, V., Dubourdieu, D.: Ecology of yeast strain saccharomyces cerevisiae during spontaneous fermentation in a bordeaux winery. Am. J. Enol. Vitic. 43(4), 375–380 (1992)

13. Kleppe, K., Ohtsuka, E., Kleppe, R., Molineux, I., Khorana, H.G.: Studies on polynucleotides: XCVI. repair replication of short synthetic DNA's as catalyzed by DNA polymerases. Journal of Molecular Biology 56(2), 341–361 (1971)

14. Kulesh, D.A., Clive, D.R., Zarlenga, D.S., Greene, J.J.: Identification of interferon-modulated proliferation-related cDNA sequences. Proceedings of the National Academy of Sciences of the United States of America 84(23), 8453–8457 (1987)

15. Malherbe, S., Fromion, V., Hilgert, N., Sablayrolles, J.-M.: Modeling the effects of assimilable nitrogen and temperature on fermentation kinetics in enological conditions. Biotechnology and Bioengineering 86(3), 261–272 (2004)

16. Mendes-Ferreira, A., del Olmo, M., Garcia-Martinez, J., Jimenez-Marti, E., Mendes-Faia, A., Perez-Ortin, J.E., Leao, C.: Transcriptional response of saccharomyces cerevisiae to different nitrogen concentrations during alcoholic fermentation. Appl. Environ. Microbiol. 73(9), 3049–3060 (2007)

17. Mendes-Ferreira, A., del Olmo, M., Garcia-Martinez, J., Jimenez-Marti, E., Leao, C., Mendes-Faia, A., Perez-Ortin, J.E.: Saccharomyces cerevisiae signature genes for predicting nitrogen deficiency during alcoholic fermentation. Applied and Environmental Microbiology 73(16), 5363–5369 (2007); PMID: 17601813 PMCID: 1950961

18. Miller, G.L.: Use of dinitrosalicylic acid reagent for determination of reducing sugar. Analytical Chemistry 31(3), 426–428 (1959)

19. Monod: La technique de culture continue; thorie et applications. Ann Ist Pasteur Lille 79, 390–410 (1950)

20. Murray, J.D.: Mathematical Biology: I. An Introduction (Interdisciplinary Applied Mathematics), 3rd edn. Springer, Heidelberg (2007)

21. Nurgel, C., Erten, H., Canbas, A., Cabaroglu, T., Selli, S.: Yeast flora during the fermentation of wines made from vitis viniferaL. cv. emir and kalecik karasi grown in anatolia. World Journal of Microbiology and Biotechnology 21(6), 1187–1194 (2005)

22. Pizarro, F., Varela, C., Martabit, C., Bruno, C., Ricardo Prez-Correa, J., Agosin, E.: Coupling kinetic expressions and metabolic networks for predicting wine fermentations. Biotechnology and Bioengineering 98(5), 986–998 (2007)

23. Rodrguez, M.E., Lopes, C.A., Barbagelata, R.J., Barda, N.B., Caballero, A.C.: Influence of candida pulcherrima patagonian strain on alcoholic fermentation behaviour and wine aroma. International Journal of Food Microbiology 138(1-2), 19–25 (2010); PMID: 20116878

24. Sainz, J., Pizarro, F., Ricardo Prez-Correa, J., Agosin, E.: Modeling of yeast metabolism and process dynamics in batch fermentation. Biotechnology and Bioengineering 81(7), 818–828 (2003)

25. Santamara, P., Garijo, P., Lpez, R., Tenorio, C., Gutirrez, A.R.: Analysis of yeast population during spontaneous alcoholic fermentation: Effect of the age of the cellar and the practice of inoculation. International Journal of Food Microbiology 103(1), 49–56 (2005)
26. Scaglia, G.J.E., Aballay, P.M., Mengual, C.A., Vallejo, M.D., Ortiz, O.A.: Improved phenomenological model for an isothermal winemaking fermentation. Food Control 20(10), 887–895 (2009)
27. Selgrade, J.F.: Dynamical behavior of a competitive model with genetic variation. Applied Mathematics Letters 2(1), 49–52 (1989)
28. Shuler, M.L., Kargi, F.: Bioprocess Engineering: Basic Concepts, 1st edn. Prentice Hall College Div. (November 1991)
29. Subramanian, V., Buck, K.K.S., Block, D.E.: Use of decision tree analysis for determination of critical enological and viticultural processing parameters in historical databases. Am. J. Enol. Vitic. 52(3), 175–184 (2001)
30. Teissier, P., Perret, B., Latrille, E., Barillere, J.M., Corrieu, G.: A hybrid recurrent neural network model for yeast production monitoring and control in a wine base medium. Journal of Biotechnology 55(3), 157–169 (1997)
31. Toro, M.E., Vazquez, F.: Fermentation behaviour of controlled mixed and sequential cultures of candida cantarellii and saccharomyces cerevisiae wine yeasts. World Journal of Microbiology and Biotechnology 18(4), 347–354 (2002)
32. Urtubia, A., Ricardo Prez-Correa, J., Soto, A., Pszczlkowski, P.: Using data mining techniques to predict industrial wine problem fermentations. Food Control 18(12), 1512–1517 (2007)
33. Vlasides, S., Ferrier, J., Block, D.: Using historical data for bioprocess optimization: modeling wine characteristics using artificial neural networks and archives process information. Biotechnology and Bioengineering 73(1), 55–68 (2001)
34. Zott, K., Miot-Sertier, C., Claisse, O., Lonvaud-Funel, A., Masneuf-Pomarede, I.: Dynamics and diversity of non-Saccharomyces yeasts during the early stages in winemaking. International Journal of Food Microbiology 125(2), 197–203 (2008); PMID: 18495281

Computational Modeling
and Verification of Signaling Pathways in Cancer

Haijun Gong[1], Paolo Zuliani[1], Anvesh Komuravelli[1],
James R. Faeder[2], and Edmund M. Clarke[1]

[1] Computer Science Department, Carnegie Mellon University
Pittsburgh, PA 15213, USA
{haijung,pzuliani,anvesh,emc}@cs.cmu.edu
[2] Department of Computational Biology, University of Pittsburgh
Pittsburgh, PA 15260, USA
faeder@pitt.edu

Abstract. We propose and analyze a rule-based model of the HMGB1
signaling pathway. The protein HMGB1 can activate a number of regulatory networks – the p53, NFκB, Ras and Rb pathways – that control
many physiological processes of the cell. HMGB1 has been recently shown
to be implicated in cancer, inflammation and other diseases. In this paper, we focus on the NFκB pathway and construct a crosstalk model of
the HMGB1-p53-NFκB-Ras-Rb network to investigate how these cou-
plings influence proliferation and apoptosis (programmed cell death) of
cancer cells. We first built a single-cell model of the HMGB1 network using the rule-based BioNetGen language. Then, we analyzed and verified
qualitative properties of the model by means of simulation and statistical
model checking. For model simulation, we used both ordinary differential equations and Gillespie's stochastic simulation algorithm. Statistical
model checking enabled us to verify our model with respect to behavioral properties expressed in temporal logic. Our analysis showed that
HMGB1-activated receptors can generate sustained oscillations of irregular amplitude for the NFκB, IκB, A20 and p53 proteins. Also, knockout
of A20 can destroy the IκB-NFκB negative feedback loop, leading to the
development of severe inflammation or cancer. Our model also predicted
that the knockout or overexpression of the IκB kinase can influence the
cancer cell's fate – apoptosis or survival – through the crosstalk of different pathways. Finally, our work shows that computational modeling
and statistical model checking can be effectively combined in the study
of biological signaling pathways.

Keywords: Model Checking, cancer, HMGB1, verification.

1 Introduction

Computational modeling is increasingly used to gain insights into the behavior of
complex biological systems, such as signaling pathways. Moreover, powerful verification methods (*e.g.*, model checking [8]) from the field of hardware verification

K. Horimoto, M. Nakatsu, and N. Popov (Eds.): ANB 2011, LNCS 6479, pp. 117–135, 2012.
© Springer-Verlag Berlin Heidelberg 2012

have been recently applied to the analysis of biological system models. In this paper we build a single-cell model of the HMGB1 pathway using the rule-based BioNetGen language [21], and use statistical model checking to formally verify interesting properties of our model. We argue that computational modeling and statistical model checking can be combined into an effective tool for analyzing the emergent behavior of complex signaling pathways. In particular, the use of statistical model checking enables us to tackle large systems in a scalable way.

The High-Mobility Group Box-1 (HMGB1) protein is released from necrotic cells or secreted by activated macrophages engulfing apoptotic cells [12]. Recent studies have shown that HMGB1 and its receptors, including the Receptor for Advanced Glycation End products (RAGEs) and Toll-Like Receptors (TLRs), are implicated in cancer, inflammation and other diseases [10,41]. Elevated expression of HMGB1 occurs in various types of tumors, including colon, pancreatic, and breast cancer [12,33,44]. HMGB1 can activate a number of regulatory networks – the PI3K/AKT, NFκB, and Ras pathways – which control many physiological processes including cell cycle arrest, apoptosis and proliferation. The cell cycle is strictly regulated and controlled by a number of signaling pathways that ensure cell proliferation occurs only when it is required by the organism as a whole [19]. Overexpression of HMGB1 can continuously activate cell-growth signaling pathways even if there are protein mutations or DNA damage, possibly leading to the occurrence of cancer in the future. Recent *in vitro* studies with pancreatic cancer cells [26] have shown that the targeted knockout or inhibition of HMGB1 and its receptor RAGE can increase apoptosis and suppress cancer cell growth. This phenomenon has also been observed with lung cancer and other types of cancer cells [4,12].

Model Checking [7,8] is one of the most widely used techniques for the automated verification and analysis of hardware and software systems. System models are usually expressed as state-transition diagrams and a temporal logic is used to describe the desired properties (specifications) of system executions. A typical property stated in temporal logic is $\mathbf{G}(grant\_req \rightarrow \mathbf{F}\,ack)$, meaning that, it is always ($\mathbf{G}$ = globally) true that a grant request eventually ($\mathbf{F}$ = future) triggers an acknowledgment. One important aspect of Model Checking is that it can be performed algorithmically – user intervention is limited to providing a system model and a property to check. Because biological systems are often probabilistic in nature, we make use of *statistical* model checking, a technique tailored to the verification of stochastic systems (see Section 2).

In [18], we proposed the first model of HMGB1 signal transduction, based on known signaling pathway studies [6,38,47]. The model was used to investigate the importance of HMGB1 in tumorigenesis. In this work, we propose a single-cell model of the HMGB1 signaling pathway, which includes the NFκB pathway and a crosstalk model of the HMGB1-p53-NFκB-Ras-Rb network. The model is described by means of the rule-based BioNetGen language [21]. We analyze and verify qualitative properties of the model using simulation and statistical model checking. For model simulation, we use both ordinary differential equations (ODEs) and Gillespie's stochastic simulation algorithm [15].

Statistical model checking enables us to verify our model with respect to behavioral properties expressed in temporal logic.

Our baseline simulations show that HMGB1-activated receptors can generate sustained oscillations of irregular amplitude for the NFκB, IκB, IKK and A20 proteins. However, mutation or knockout of the A20 protein can destroy the IκB-NFκB negative feedback loop, leading to the development of severe inflammation or cancer. Further analysis shows that overexpression of HMGB1 can up-regulate the oncoproteins NFκB and Cyclin E (which regulate cell proliferation), but down-regulate the tumor-suppressor protein p53 (which regulates cell apoptosis). Also, overexpression of NFκB can increase the expression level of both Cyclin E and p53. Our model also predicts that the knockout or overexpression of the IκB kinase (IKK) can influence the cancer cell's fate – apoptosis or survival – through the crosstalk of different pathways. To the best of the authors' knowledge, this work is the first attempt to integrate the NFκB, p53, Ras, and Rb signaling pathways activated by HMGB1 in one rule-based model.

2 Statistical Model Checking

In the past few years, there has been growing interest in the formal verification of stochastic systems, and biological systems in particular [25,28,39], by means of model checking techniques. The verification problem is to decide whether a stochastic model satisfies a temporal logic property with a *probability* greater than or equal to a certain threshold. To express properties, we use a temporal logic in which the temporal operators are equipped with *bounds*. For example, the property "p53 will always stay below 30 in the next 80 time units" is written as $\mathbf{G}^{80}(p53 < 30)$. We ask whether our stochastic system M satisfies that formula with a probability greater than or equal to a fixed threshold (say 0.99), and we write $M \models Pr_{\geqslant 0.99}[\mathbf{G}^{80}(p53 < 30)]$. Such questions can be answered by *Statistical Model Checking* [50], the technique we use for verifying BioNetGen models simulated by Gillespie's algorithm.

Statistical model checking treats the verification problem as a statistical inference problem and solves it by randomized sampling of traces (simulations) from the model. In particular, the inference problem can be solved by means of hypothesis testing or estimation. The former amounts to deciding between two hypotheses – $M \models Pr_{\geqslant \theta}[\phi]$ versus $M \models Pr_{<\theta}[\phi]$, where θ is a given probability threshold and ϕ is a temporal logic property. The latter, instead, approximates probabilistically (that is, it computes with high probability an *estimate* close to) the true probability p that ϕ holds, and then compares that estimate with θ. In both approaches, sampled traces are model checked individually to determine whether property ϕ holds, and the number of satisfying traces is used by the hypothesis testing (or estimation) procedure to decide between $p \geqslant \theta$ and $p < \theta$. (In the case of estimation, one also has an estimate that is close to p with high probability.) Note that statistical model checking cannot guarantee a correct answer to the verification problem. However, the probability of giving a wrong answer can be arbitrarily bounded by the user.

In the next section we describe the temporal logic used in this work, Bounded Linear Temporal Logic (BLTL) [25,51].

2.1   Bounded Linear Temporal Logic

Let SV be a finite set of real-valued variables. An atomic proposition AP is a boolean predicate of the form $e_1 \sim e_2$, where e_1 and e_2 are arithmethic expressions over variables in SV, and $\sim$ is either $\geq$, $\leq$, or $=$. A BLTL property is built over atomic propositions using boolean connectives and bounded temporal operators. The syntax of the logic is the following:

$$\phi ::= AP \mid \phi_1 \vee \phi_2 \mid \phi_1 \wedge \phi_2 \mid \neg\phi_1 \mid \phi_1 \mathbf{U}^t \phi_2.$$

The bounded until operator $\phi_1 \mathbf{U}^t \phi_2$ requires that, *within* time t, ϕ_2 will be true and ϕ_1 will hold until then. Bounded versions of the $\mathbf{F}$ and $\mathbf{G}$ operators can be easily defined: $\mathbf{F}^t \phi = true \ \mathbf{U}^t \phi$ requires ϕ to hold true within time t; $\mathbf{G}^t \phi = \neg\mathbf{F}^t \neg\phi$ requires ϕ to hold true up to time t.

The semantics of BLTL is defined with respect to *traces* (or executions) of a system. In our case, a trace will be the output of a BioNetGen model simulated by Gillespie's algorithm. Formally, a trace is a sequence of time-stamped state transitions of the form $\sigma = (s_0, t_0), (s_1, t_1), ...$, where (s_i, t_i) denotes that the system moved to state s_{i+1} after having sojourned for t_i time units in state s_i. The fact that a trace σ satisfies the BLTL property ϕ is written $\sigma \models \phi$. We denote the trace suffix starting at step k by σ^k.

We have the following semantics of BLTL:

1. Note that the semantics of BLTL is defined over *infinite* traces, while of course any simulation trace must be finite in length.
2. It can be shown that traces of an appropriate (finite) length are sufficient to decide BLTL properties.

The interested reader can find details elsewhere [51].

2.2   Bayesian Statistical Model Checking

We recently introduced sequential Bayesian hypothesis testing and estimation techniques and applied them to the verification of signaling pathways and other stochastic systems [25,51]. Sequential sampling means that the number of sampled traces is not fixed a priori, but it is instead determined at "run-time," depending on the evidence gathered by the samples seen so far. This often leads to a significantly smaller number of sampled traces. Both approaches are based on Bayes' theorem, which enables us to use prior information about the model being verified, where available. We now briefly describe both techniques.

Bayesian Hypothesis Testing. The hypothesis test is based on the Bayes Factor, which is the likelihood ratio of the sampled data with respect to the two hypotheses. For statistical model checking, the hypotheses being tested are $H_0 : p \geqslant \theta$ and $H_1 : p < \theta$, where p is the (unknown) probability that our model satisfies a given property, and θ is a probability threshold. Formally, the Bayes Factor of data d and hypotheses H_0 and H_1 is $B = \frac{Pr(d|H_0)}{Pr(d|H_1)}$. Therefore, B can be interpreted as a measure of evidence (given by the data d) in favor of H_0. Now, fix an evidence threshold $T > 1$. Our algorithm iteratively draws independent and identically distributed (iid) sample traces $\sigma_1, \sigma_2, ...$, and checks whether they satisfy ϕ. After each trace, the algorithm computes the Bayes Factor B to check if it has obtained conclusive evidence. The algorithm accepts H_0 if $B > T$, and rejects H_0 (accepting H_1) if $B < \frac{1}{T}$. Otherwise (if $\frac{1}{T} \leqslant B \leqslant T$), it continues drawing iid samples. It can be shown that when the algorithm terminates, the probability of a wrong answer is bounded above by $\frac{1}{T}$. The algorithm is shown below in Algorithm 1 – full details can be found elsewhere [51].

Algorithm 1. Statistical Model Checking by Bayesian Hypothesis Testing
Require: BLTL Property ϕ, Probability threshold $\theta \in (0, 1)$, Threshold $T > 1$, Prior density g for unknown parameter p

$n := 0$ *{number of traces drawn so far}*
$x := 0$ *{number of traces satisfying ϕ so far}*
loop
    $\sigma :=$ draw a sample trace of the system (iid)
    $n := n + 1$
    if $\sigma \models \phi$ **then**
        $x := x + 1$
    end if
    $B :=$ BayesFactor(n, x) *{compute the Bayes Factor}*
    if $(B > T)$ **then**
        return H_0 accepted
    else if $(B < \frac{1}{T})$ **then**
        return H_1 accepted
    end if
end loop

Bayesian Interval Estimation. Recall that in estimation, we are interested in computing a value (an estimate) which is close to p with high probability, the true probability that the model satisfies the property. The estimate is usually in the form of a confidence interval – an interval in $[0, 1]$ which contains p with high probability. Our estimation method follows directly from Bayes' theorem. Given a prior distribution over p and sampled data, Bayes' theorem enables us to obtain the *posterior* distribution of p (*i.e.*, the distribution of p given the data sampled and the prior). This means that we can estimate p with the mean of the posterior distribution. Furthermore, by integrating the posterior over a suitably chosen interval, we can compute a Bayes interval estimate with any given confidence coefficient. Fix a confidence $c \in (\frac{1}{2}, 1)$ and a half-width $\delta \in (0, \frac{1}{2})$. Our algorithm iteratively draws iid traces, checks whether they satisfy ϕ, and builds an interval of total width 2δ, centered on the posterior mean. If the integral of the posterior over this interval is greater than c, the algorithm stops; otherwise, it continues sampling. The algorithm is given in Algorithm 2. Again, full details are given in [51].

Algorithm 2. Statistical Model Checking by Bayesian Interval Estimates

Require: BLTL Property ϕ, half-interval size $\delta \in (0, \frac{1}{2})$, interval coefficient $c \in (\frac{1}{2}, 1)$, Prior Beta distribution with parameters α, β

$n := 0$ *{number of traces drawn so far}*
$x := 0$ *{number of traces satisfying ϕ so far}*
repeat
    $\sigma :=$ draw a sample trace of the system (iid)
    $n := n + 1$
    if $\sigma \models \phi$ **then**
        $x := x + 1$
    end if
    $\hat{p} := (x + \alpha)/(n + \alpha + \beta)$ *{compute posterior mean}*
    $(t_0, t_1) := (\hat{p} - \delta, \hat{p} + \delta)$ *{compute interval estimate}*
    if $t_1 > 1$ **then**
        $(t_0, t_1) := (1 - 2 \cdot \delta, 1)$
    else if $t_0 < 0$ **then**
        $(t_0, t_1) := (0, 2 \cdot \delta)$
    end if
    {compute posterior probability of $p \in (t_0, t_1)$}
    $\gamma := \text{PosteriorProb}(t_0, t_1)$
until $(\gamma \geqslant c)$
return $(t_0, t_1), \hat{p}$

3 Crosstalk Model of HMGB1

Apoptosis and cell proliferation are two important processes in cancer and are respectively regulated by two proteins – p53 and Cyclin E – acting in two different signaling pathways. The protein p53 is a tumor suppressor whose activation can lead to cell cycle arrest, DNA repair or apoptosis. Cyclin E is a cell cycle regulatory protein that regulates the G1-S phase transition during cell proliferation. The behavior of these two signaling pathways can be influenced by crosstalk or coupling with other pathways and proteins.

3.1 Motivations

Experimental studies have shown that HMGB1 can activate three fundamental downstream signaling pathways: the PI3K/AKT, RAS-ERK and NFκB pathways. These in turn lead to the activation of two other signaling pathways: the p53-MDM2 and Rb-E2F pathways, which regulate apoptosis and cell proliferation, respectively. In [18], we proposed the first computational model for HMGB1 signal transduction (also called the NFκB-knockout model). The model included the p53-MDM2, Ras-ERK, and Rb-E2F pathways and was able to explain qualitatively some existing experimental phenomena in tumorigenesis. One of our goals in this work is to integrate the NFκB signaling pathway into our previous NFκB-knockout model in order to explain recent results linking overexpression of HMGB1 with a decrease of apoptosis (and increased cancer cell survival).

The NFκB protein is involved in a variety of cellular processes, including inflammation, cell proliferation and apoptosis. Studies have shown that NFκB is also a transcription factor for the pro-apoptotic gene p53 [48], for anti-apoptotic genes, including Bcl-XL [23] and for the cell-cycle regulatory proteins Myc and Cyclin D [20]. We aim to understand how the NFκB pathway influences the HMGB1 signal transduction pathway.

Recent experiments with mammalian cells [22,36] have found oscillations of NFκB, activated by tumor necrosis factor (TNF), with a time period in the order of hours. Several mathematical models based on ODEs were constructed to study the NFκB system [22,27,32]. Since biological systems are intrinsically stochastic, our goal is to study the oscillations of NFκB's expression level in the nucleus and compare the stochastic simulation results with the ODEs results.

Finally, the NFκB pathway is regulated by many proteins including A20, IKK and NFκB. The overexpression or mutation of IKK and NFκB [5,11] occur frequently in many cancer types. We aim to investigate how these proteins' mutation or overexpression changes the cell's fate – apoptosis or survival.

3.2 Model Formulation

In Fig. 1, we illustrate the crosstalk model of the HMGB1 signaling pathway. It includes 44 molecular species (nodes), 82 chemical reactions, and four coupling signaling pathways: the RAS-ERK, Rb-E2F, IKK-NFκB and p53-MDM2 pathways. We now briefly describe these signaling pathways and their interplay with the NFκB network. We denote activation (or promotion) by $\rightarrow$ and inhibition (or repression) by $\dashv$.

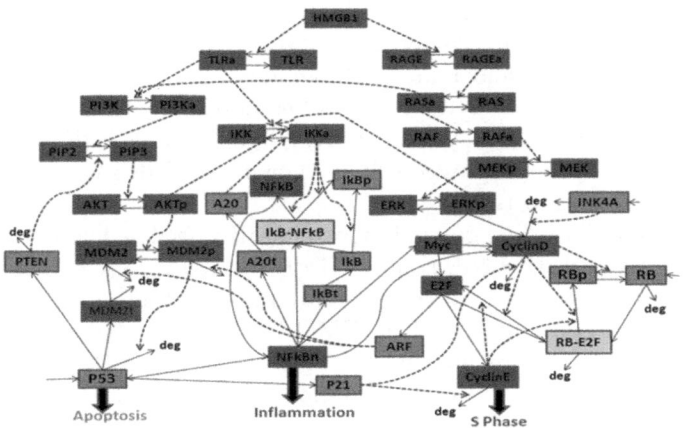

Fig. 1. Schematic view of HMGB1 signal transduction. Blue nodes represent tumor suppressor proteins; red nodes represent oncoproteins/lipids; brown nodes represent protein complexes. Solid lines with arrows denote protein transcription, degradation or changes of molecular species; dashed lines with arrows denote activation processes.

The p53-MDM2 pathway is regulated by a negative feedback loop: TLR → PI3K → PIP3 → AKT → MDM2 ⊣ p53 → MDM2, and a positive feedback loop: p53 → PTEN ⊣ PIP3 → AKT → MDM2 ⊣ p53 → Apoptosis [29]. The protein PI3K is activated by the toll-like receptors (TLR2/4) [45] and can phosphorylate the lipid PIP2 to PIP3, leading to the phosphorylation of AKT. The oncoprotein MDM2 can only reside in the cytoplasm before it is phosphorylated by AKT. The phosphorylated MDM2 can enter the nucleus to bind with p53, inhibit p53's transcriptional activity and target it for degradation. The protein p53 can also induce the transcription of another tumor suppressor protein, PTEN, which can hydrolyze PIP3 to PIP2 and inhibit the phosphorylation of MDM2.

The RAS-ERK pathway is RAGE → RAS → RAF → MEK → ERK. Upon activation by HMGB1, RAGE will activate the RAS proteins, leading to a cascade of events including the activation and phosphorylation of the RAF, MEK and ERK1/2 proteins. The mutated K-RAS protein, a member of the RAS protein family, can continuously activate the downstream cell cycle signaling pathways. The activated ERK can enter the nucleus and phosphorylate transcription factors which induce the expression of cell cycle regulatory proteins, such as Cyclin D and Myc (see Fig. 1).

The Rb-E2F pathway is Cyclin D ⊣ Rb ⊣ E2F → Cyclin E ⊣ Rb. This pathway plays an important role in the regulation of the G1-S phase transition in the cell cycle. In particular, E2F is a transcription factor that regulates the expression of a set of cell-cycle regulatory genes [49]. In resting cells, E2F's transcriptional activity is repressed by the unphosphorylated Rb, a tumor suppressor protein, through the formation of an Rb-E2F complex. The oncoproteins Cyclin D and Myc can phosphorylate the Rb protein, which can then activate E2F. In turn, E2F activates the transcription of Cyclin E and Cyclin-dependent protein kinase 2 (CDK2), which promotes cell-cycle progression from G1 to S phase. Cyclin E can also phosphorylate and inhibit Rb, leading to a forward positive feedback loop [42,37]. The protein INK4A is another important tumor suppressor that can repress the activity of Cyclin D-CDK4/6 and inhibit E2F's transcriptional activity and cell cycle progression. It is known that INK4A is mutated in over 90% of pancreatic cancers [3].

The NFκB pathway is regulated by two negative feedback loops: TLR → IKK ⊣ IκB ⊣ NFκB → IκB ⊣ NFκB, and NFκB → A20 ⊣ IKK ⊣ IκB ⊣ NFκB. In the resting wild-type cells, IκB resides only in the cytoplasm where it is bound to NFκB. Upon being activated by HMGB1, TLR2/4 can signal via MyD88, IRAKs and TRAF to activate and transform IκB kinase (IKK) into its active form IKKa, leading to the phosphorylation, ubiquitination and degradation of IκB. The free NFκB rapidly enters the nucleus to bind to specific κB sites in the A20 and IκB promoters, activating their expression. The newly synthesized IκB enters the nucleus to bind to NFκB and takes it out into the cytoplasm to inhibit its transcriptional activity. Moreover, the newly synthesized A20 can also inhibit IKK's activity, leading to inhibition of NFκB.

Besides the main signal transduction, the interplay between these four signaling pathways can influence the cell's fate. As shown in Fig. 1, RAS can activate the PI3K-AKT signaling pathway; ERK and AKT can activate IKK in the NFκB pathway. The tumor suppressor protein ARF, activated by the overexpressed on-coprotein E2F, can bind to MDM2 to promote its degradation and stabilize p53's expression level, leading to apoptosis. Moreover, it has been demonstrated [46] that the p53-dependent tumor suppressor proteins p21 and FBXW7 can restrain the activity of Cyclin D-CDK4/6 and Cyclin E-CDK2 (only p21 is shown in Fig. 1 to represent both p21 and FBXW7's contribution). Mutations of RAS, ARF, P21 and FBXW7 have been found in many cancers [3,9]. NFκB is a transcription factor for p53, Myc and Cyclin D, regulating cell proliferation and apoptosis. The over-expression of NFκB occurs in approximately 80% of lung cancer cases [43], and it is also common in pancreatic cancer [5]. Our model and simulation will investigate how these mutations and overexpressions affect the cell's fate.

3.3   Simulation Models

Similar to the model in our previous work [18], in this model (see Fig. 1), all sub-strates are expressed in terms of the number of molecules. A protein with the subscript "a", "p" or "t" corresponds respectively to active form, phosphorylated form or mRNA transcript of the protein. For example:

- AKT (AKT$_p$) - unphosphorylated (phosphorylated) AKT.
- RAS (RAS$_a$) - inactive (active) RAS.
- IκB$_t$ - mRNA transcript of IκB.

We sometimes use CD to stand for the Cyclin D-CDK4/6 complex, CE for the Cy-clin E-CDK2 complex, RE for the Rb-E2F complex, and IκNF for the (IκB|NFκB) or (IκB-NFκB) dimer. We also assume that the total number of active and inac-tive forms of the RAGE, TLR, PI3K, IKK, PIP, AKT, RAS, RAF, MEK, ERK and NFκB molecules is constant [18]. For example, AKT + AKT$_p$ = AKT$_{tot}$, PIP2 + PIP3 = PIP$_{tot}$ and NFκB + NFκB$_n$ + (IκB|NFκB) = NFκB$_{tot}$.

We have formulated a reaction model corresponding to the reactions illustrated in Fig. 1 in the form of rules specified in the BioNetGen language [21]. We use Hill functions to describe the rate laws governing the transcription of some pro-teins, including PTEN, MDM2, CyclinD (CD), Myc, E2F, CyclinE (CE), A20 and IκB, and use mass action rules for other reactions. We use both ODEs and Gillespie's stochastic simulation algorithm (SSA) [15] to simulate the same model with BioNetGen. Stochastic simulation is important because when the number of molecules involved in the reactions is small, stochasticity and discretization effects become more prominent [17,16,31]. The ODEs for the NFκB-knockout HMGB1 model have been provided in our previous work [18]. The ODEs for the HMGB1-p53-Ras-NFκB-Rb crosstalk model are listed in the online supplementary materi-als [2]. We now give an example to illustrate how to convert an ODE into BioNet-Gen rules. The ODE for the phosphorylated $AKT - AKT_p$ is

$$\frac{d}{dt}AKT_p(t) = k_4 PIP3(t)AKT(t) - d_4 AKT_p(t),$$

where the first term describes the phosphorylation of AKT, activated by PIP3. The second term describes AKT_p dephosphorylation. In BioNetGen, the molecule type $AKT(a \sim U \sim p)$ has a component named a with state label U (unphosphorylated) and p (phosphorylated). The BioNetGen rules for the ODE above are:

$$AKT(a \sim U) + PIP3 \to AKT(a \sim p) + PIP3 \quad k4$$

$$AKT(a \sim p) \to AKT(a \sim U) \quad d4$$

where k_4 and d_4 are the constants for AKT phosphorylation and dephosphorylation rates, respectively. The interested reader can refer to the BioNetGen tutorial [13] for details. The BioNetGen code of our model is available online [1]. The model contains a large number of undetermined parameters which are difficult to estimate from available experimental data or from the literature. We emphasize that in this work, the values for several undetermined parameters listed in [2] have been chosen in order to produce a qualitative agreement with previous experiments.

4   Simulation Results

To validate the properties of the HMGB1 signal transduction model, we have conducted a series of deterministic and stochastic simulations and compared our results with known experimental facts. In our model, the p53-MDM2 and NFκB signaling pathways are regulated by two feedback loops. Recent experimental results have shown that p53's and MDM2$_p$'s expression levels undergo oscillations in response to stress signals. For example, oscillations lasted more than 72 hours after γ irradiation in Geva-Zatorsky et al.'s experiment [14]. Also, Hoffmann's experiment found oscillations of NFκB in response to TNF stimulation, with four equally spaced peaks over the course of the 6-hour experiment [22]. We first conducted baseline simulations for several important proteins involved in the HMGB1 signaling pathway. In our simulations, we set the initial value for the number of HMGB1 molecules to be 10^2; the nonzero initial values for other proteins are listed in Table 1. The input parameters and reaction descriptions are listed in the online supplementary materials [2].

Table 1. Initial values for the proteins in the crosstalk model of HMGB1

Proteins	TLR	PI3K	PIP2	AKT	MDM2	MDM2$_p$	p53	IκB-NFκB
# of Mol.	10^3	10^5	10^5	10^5	10^4	2×10^4	2×10^4	10^5
Proteins	RAGE	RAS	RAF	MEK	ERK	RE	IKK	
# of Mol.	10^3	10^4	10^4	10^4	10^4	10^4	10^5	

In Fig. 2, we give the dynamic of the NFκB, IKK, IκB-NFκB complex, IκB, and A20 proteins using both stochastic simulation and ODEs. In Fig. 2 (A,D), we see that IKK, upon being stimulated by HMGB1, is activated immediately by the TLR, AKT and ERK proteins. This leads to the phosphorylation of IκB isoform (Fig. 2 B,E), which in turn allows NFκB to translocate into the nucleus.

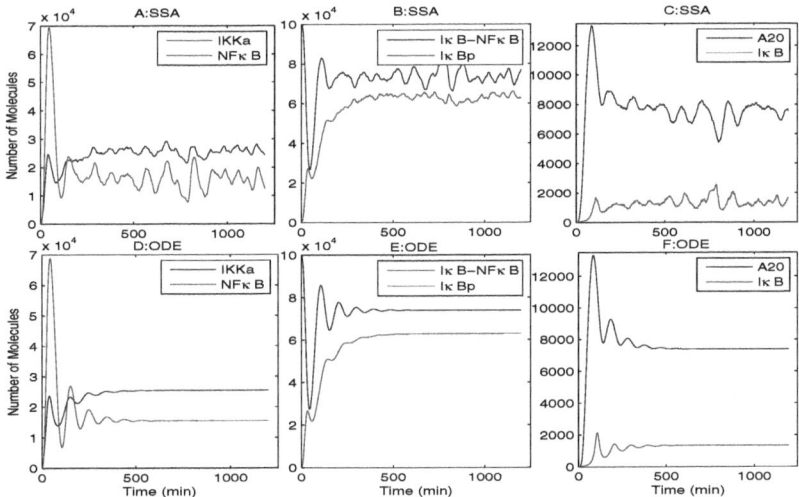

Fig. 2. Number of IKKa, NFκB (A,D), IκB$_p$, IκB-NFκB complex (B,E), IκB and A20 (C,F) molecules versus time for baseline simulations with SSA(A-C) and ODE(D-F) models

There, NFκB binds to the DNA and induces the transcription of the IκB and A20 inhibitor genes (Fig. 2 C,F). The synthesized IκB can enter the nucleus and recapture NFκB back into the cytoplasm to form the IκB-NFκB complex. However, IκB is continuously phosphorylated and degraded, resulting in the continued translocation of NFκB. The stochastic simulation of HMGB1-induced NFκB oscillation depicted in Fig. 2A fits very well with Nelson's experimental results [36] – the oscillation of NFκB continued for more than 20 hours after continuous TNFα stimulation, damping slowly with a period of 60-100 minutes. However, the ODEs simulation results in Fig. 2D show no NFκB oscillation after 500 minutes, when the cell reaches the resting state. The phosphorylation of IκB leads to the decrease of the IκB-NFκB complex in Fig. 2(B,E). The A20 protein can inactivate IKK to stabilize the IκB-NFκB complex. The stochastic simulation shows continuous oscillation (Fig. 2C) in 20 hours, but no oscillation is present in the ODE simulation (Fig. 2F) after 400 minutes. This discrepancy shows that in modeling signal transduction, it is important to capture accurately both the discretization and stochasticity of chemical reactions. Similar stochastic oscillations of p53 and MDM2 proteins are shown in the online supplementary materials [2].

The A20 protein plays an important role in the regulation of the NFκB network. It is known that A20 knockout can result in severe inflammation and tissue damage in multiple organs [24]. As Fig. 3 shows, when A20 is knocked out, over 90% of IKK is activated, which can then phosphorylate and ubiquitinate IκB. This leads to the disassembly of the IκB-NFκB dimer and liberation of NFκB, which rapidly translocates into the nucleus. The A20-knockout results in Fig. 3 demonstrate that the oscillation of NFκB dampens very quickly, with a small period compared to Fig. 2A. This phenomenon is consistent with Mengel et al.'s discovery that A20

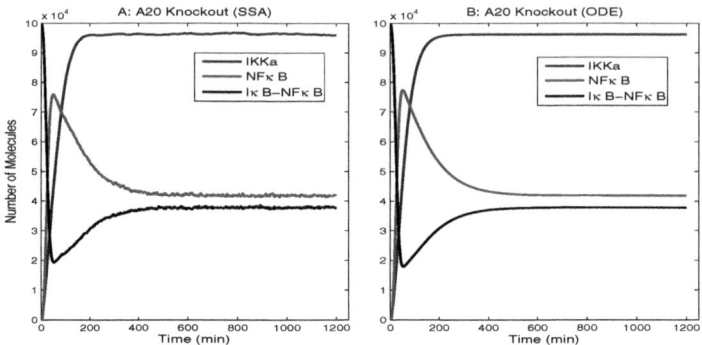

Fig. 3. Number of IKKa, NFκB, IκB-NFκB molecules versus time in the A20-knockout model

can not only dampen the oscillations, but also control the oscillation period of NFκB [35]. So, the loss of A20 can destroy the IκB-NFκB negative feedback loop. The precise role of the A20 negative feedback remains to be elucidated in future experiments.

A number of studies have found that overexpression of HMGB1 and its receptors is associated with cancer [12,33]. Our recent NFκB-knockout HMGB1 model [18] qualitatively explained the experimental result that overexpression of HMGB1 decreases apoptosis and promotes DNA replication and proliferation in cancer cells. We now ask the following question: How do the expression levels of HMGB1 and other proteins influence the cell's fate when the NFκB signaling pathway is integrated?

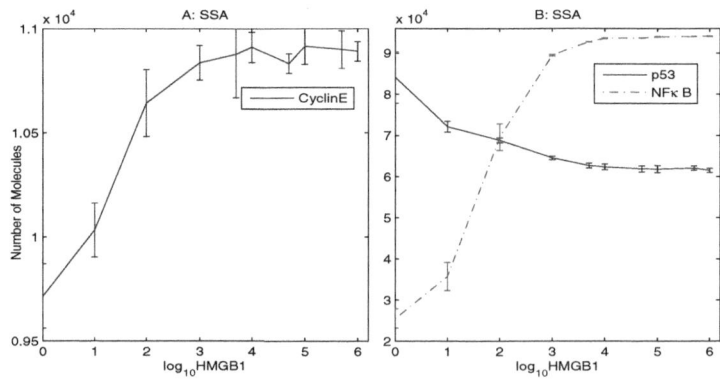

Fig. 4. Overexpression of HMGB1 leads to the increase of DNA replication proteins Cyclin E and Nuclear Factor NFκB and the decrease of p53 with the SSA model

In Fig. 4, we varied the level of HMGB1 to determine how it affects cell behavior. We increased the number of HMGB1 molecules from 1 to 10^6, and Cyclin E's expression level at 300 minutes, and the first maximum of p53 and NFκB in phase G1 were measured using stochastic simulation. All the experiments were repeated 10 times per value to compute the mean and standard errors. In Fig. 4, we see that an increase of HMGB1's initial value can increase the number of Cyclin E and NFκB molecules, and decrease p53's expression level. With respect to our previous model [18], we see that the expression level of Cyclin E and p53 are higher, since NFκB can induce the transcription of p53, Cyclin D and Myc, which can activate the expression of Cyclin E during cell cycle progression. Therefore, the knockout of HMGB1 and its receptors can inhibit the expression of NFκB and Cyclin E, leading to cell cycle arrest or inhibition of cancer cell proliferation.

The expression of the IKK protein is elevated in many cancer cells [11]. Since IKK regulates NFκB's DNA-binding activity, we investigated how the dynamic of IKK influences the expression levels of the cell-cycle regulatory proteins Cyclin E and NFκB. We increased the number of IKK molecules and measured Cyclin E's expression level at 300 minutes. As for NFκB, we measured two values: the first maximum and the expression level at 300 minutes. In Fig. 5(A,B) we see that with the increase of IKK's expression level, Cyclin E and NFκB's concentrations increase quickly, since more active IKK can promote NFκB's DNA-binding and transcriptional activity, accelerating the progression of cell proliferation or inflammation. It has been observed that NFκB plays a key role in the development and progression of cancer, including proliferation, migration and apoptosis [5]. Aberrant or constitutive NFκB activation has been detected in many cancers [5,43]. Furthermore, overexpression of NFκB is very common in pancreatic cancer [5]. In our model, we set the initial value for NFκB to 0, so that NFκB is only found in the form of the transient IκB-NFκB dimer. In Fig. 5C, we increased the initial value of IκB-NFκB dimers and measured the pro-apoptotic protein p53 and cell-cycle regulatory protein Cyclin E's expression level. The results demonstrate that the overexpression of NFκB can increase Cyclin E's concentration, thereby promoting cancer cell proliferation. However, for the pro-apoptotic protein p53,

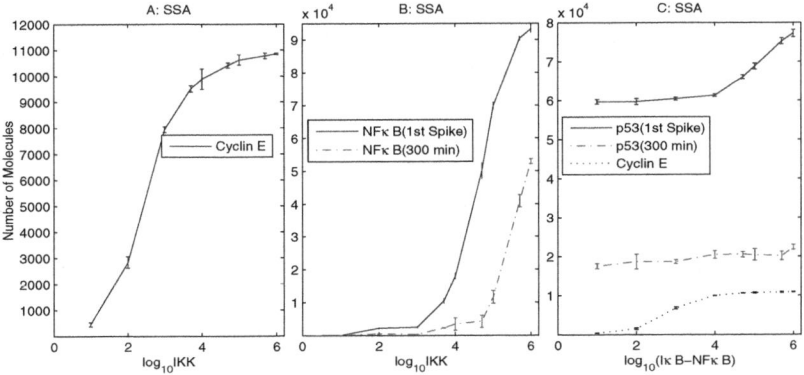

Fig. 5. Overexpression of IKK leads to the increase of Cyclin E and NFκB (A-B); overexpression of NFκB increases Cyclin E and p53's concentration (C)

the simulations show that the amplitude of p53's first maximum increases sharply when the number of NFκB-IκB dimers is over 10^4. The expression level at 300 minutes (in the steady state) is almost stable even when the number of NFκB-IκB dimers reach 10^6. This is because p53's expression level is regulated by its negative regulator MDM2 and stays at a low level in the resting state. Fig. 5 explains the experimental discovery that the overexpression of IKK and NFκB decreases apoptosis and promotes DNA replication and proliferation in cancer cells [5,11] (though NFκB could also induce the transcription of p53).

The results visualized in Fig. 4 and Fig. 5 provide some ways to inhibit tumor cell proliferation and induce tumor cell apoptosis through inhibition or deactivation of the HMGB1 and NFκB signaling pathways. This can be achieved, for example, via the inhibition of IKK and NFκB's transcription activity on Cyclin D and Myc. Recently, the targeting of IKK and IKK-related kinases has become a popular avenue for therapeutic interventions in cancer [30]. Inhibitor drugs for NFκB's upstream protein RAS [34,40], and downstream protein CDK, have also been developed to inhibit tumor growth.

5   Verification of the HMBG1 Pathway

We applied statistical model checking to formally verify several important properties related to NFκB. We first applied the Bayesian Hypothesis Testing method to verify the properties in the stochastic HMGB1 model. We tested whether our model satisfied a given BLTL property with probability $p \geqslant 0.9$. We set the Bayesian Hypothesis Testing threshold $T = 1000$, so the probability of a wrong answer was smaller than 10^{-3}.

Property 1: It is known that NFκB in the nucleus increases quickly after IκB is phosphorylated by IKK, which is activated by HMGB1 after approximately 30-60 minutes. Let $R = \frac{NF\kappa B_n}{NF\kappa B_{tot}}$ be the fraction of NFκB molecules in the nucleus. We verified the following property

$$Pr_{\geqslant 0.9}[\mathbf{F}^t(R \geqslant a)],$$

which informally means that the fraction of NFκB molecules in the nucleus will eventually be greater than a threshold value a within t minutes. We verified this property with various values of a and t. The results are shown in Table 2.

Table 2. Verification of Property 1 $(HMGB1 = 10^2)$

Property: $Pr_{\geqslant 0.9}[\mathbf{F}^t(NF\kappa B_n/NF\kappa B_{tot} \geqslant a)]$

t(min)	a	# of Samples	# of Successes	Result	Time (s)
30	0.4	22	22	True	21.30
30	0.45	92	87	True	92.86
30	0.5	289	45	False	537.74
60	0.65	22	22	True	26.76

Property 2: The IκB and A20 proteins, which are NFκB's transcription targets, inhibit the expression of NFκB, leading to the oscillation of NFκB's expression level. We verified the property

$$Pr_{\geqslant 0.9}[\mathbf{F}^t(R \geqslant 0.65 \ \& \ \mathbf{F}^t(R \leqslant 0.20 \ \& \ \mathbf{F}^t(R \geqslant 0.20 \ \& \ \mathbf{F}^t(R \leqslant 0.20))))] \ .$$

That is, the fraction of NFκB molecules in the nucleus is oscillating: R will eventually be greater than 65% within t minutes, it will then fall below 20% within another t minutes, will increase over 20% within the following t minutes, and will finally decrease to 20% within another t minutes. We verified this property with various values of t and HMGB1, and the results are shown in Table 3.

Table 3. Verification of Property 2

Property: $Pr_{\geqslant 0.9}[\mathbf{F}^t(R \geqslant 0.65 \ \& \ \mathbf{F}^t(R \leqslant 0.20 \ \& \ \mathbf{F}^t(R \geqslant 0.20 \ \& \ \mathbf{F}^t(R \leqslant 0.20))))]$

HMGB1	t(min)	# of Samples	# of Successes	Result	Time (s)
10^2	45	13	1	False	76.77
10^2	60	22	22	True	111.76
10^2	75	104	98	True	728.65
10^5	30	4	0	False	5.76

Property 3: A large proportion of PI3K, RAS and IKK molecules can be activated when the overexpressed HMGB1 binds to RAGE and TLRs. We verified the following property

$$Pr_{\geqslant 0.9}[\mathbf{F}^t\mathbf{G}^{180}(PI3K_a/PI3K_{tot} > 0.9 \ \& \ RAS_p/RAS_{tot} > 0.8 \ \& \ IKK_a/IKK_{tot} > 0.6)],$$

which means that 90% of PI3K, 80% of RAS and 60% of IKK will be activated within t minutes, and they will always stay above these values during the next 3 hours. This property was tested with HMGB1 overexpressed (10^5) and for various values of t given in Table 4.

Table 4. Verification of Property 3 and 4

Property 3					Property 4				
t(min)	Samples	Successes	Result	Time (s)	IKK	Samples	Successes	Result	Time (s)
90	9	0	False	21.27	10^5	22	22	True	547.52
110	38	37	True	362.19	2×10^4	9	2	False	55.86
120	22	22	True	214.38	10^2	4	0	False	16.89

Property 4: The overexpression of IKK can promote the translocation of NFκB into the nucleus, induce the transcription of protein Cyclin D and Myc and lead to the overexpression of Cyclin E. We verified the property

$$Pr_{\geqslant 0.9}[\mathbf{F}^{300}\mathbf{G}^{300}(CyclinE >= 10,000)].$$

The results are presented in Table 4.

We also used the Bayesian interval estimation algorithm to perform a more accurate study of several temporal properties. In Table 5, we report the estimates for the probability that the HMGB1-NFκB model satisfies three temporal logic properties. We ran the tests with uniform prior and half-interval size $\delta = 0.01$ and coverage probability $c = 0.9$. We can see from the computation time of the tables that statistical model checking is feasible even with large reaction networks, such as the one under study.

Table 5. Bayesian Estimation of Temporal Logic Properties

IKK	Property	Posterior Mean	# of Samples	Time (s)
10^5	$[\mathbf{F}^{30}(NF\kappa B_n/NF\kappa B_{tot} \geqslant 0.45)]$	0.9646	903	464
10^5	$[\mathbf{F}^{60}(NF\kappa B_n/NF\kappa B_{tot} \geqslant 0.65$ &			
	$\mathbf{F}^{60}(NF\kappa B_n/NF\kappa B_{tot} \leqslant 0.2))]$	0.9363	689	1783
10^2	$[\mathbf{F}^{300}\mathbf{G}^{300}(CyclinE >= 10,000)]$	0.0087	113	252.83

6   Discussion

This paper is the first attempt to integrate the NFκB signaling pathway with the p53-MDM2 and Rb-E2F pathways to study HMGB1 signal transduction at the single cell level. The NFκB pathway is important because it regulates the transcription of many pro-apoptotic and anti-apoptotic proteins. Several experiments were simulated using ODEs and Gillespie's algorithm under a range of conditions, using the BioNetGen language and simulator. We used statistical model checking to formally and automatically validate our model with respect to a selection of temporal properties. Model validation is performed efficiently and in a scalable way, thereby promising to be feasible even for larger BioNetGen models.

Our stochastic simulations show that HMGB1-activated receptors can generate sustained oscillations of irregular amplitude for several proteins including NFκB, IKK and p53. These results are qualitatively confirmed by experiments on p53 [14] and NFκB [36]. The simulations also demonstrate a dose-dependent p53, Cyclin E and NFκB response curve to an increase in HMGB1 stimulus, which is qualitatively consistent with experimental observations in cancer studies [26,44]. In particular, overexpression of HMGB1 can promote the expression of the cell cycle regulatory proteins Cyclin E and NFκB. It can also inhibit the pro-apoptotic p53 protein, which can lead to increased cancer cell survival and decreased apoptosis. We also investigated how the mutation or knockout of the IKK, A20 and NFκB proteins influence the fate of cancer cells.

Moreover, understanding of HMGB1 at the mechanistic level is still not clear, and reaction rates for some proteins interactions in the four signaling pathways have not been measured by experiments. We have also made some simplifications and assumption in our model. For example, the NFκB protein complex is composed of RelA(p65), RelB, cRel and NFκB1(p50), but we neglected the formation of the NFκB complex in our HMGB1 model in order to make the model relatively simple.

Our current HMGB1-Ras-p53-NFκB-Rb crosstalk model compares qualitatively well with experiments, and can provide valuable information about the behavior of HMGB1 signal transduction in response to different stimuli. In the future we plan to improve further our model with the help of new experimental results. Furthermore, the use of model checking techniques will enable us identifying and validating more realistic models.

Acknowledgments. This work was supported by a grant from the U.S. National Science Foundation's Expeditions in Computing Program (award ID 0926181). The authors thank Michael T. Lotze (University of Pittsburgh) for calling their attention to HMGB1 and for helpful discussions on the topic. H.G. would like to thank Marco E. Bianchi (San Raffaele University) for email discussions on HMGB1. The authors would also like to thank Ilya Korsunsky and Máté L. Nagy for their comments on this paper.

References

1. HMGB1-NFkB BioNetGen Code,
   http://www.cs.cmu.edu/~haijung/research/HMGB1ANB.bngl
2. Online Supplementary Materials,
   http://www.cs.cmu.edu/~haijung/research/ANBSupplement.pdf
3. Bardeesy, N., DePinho, R.A.: Pancreatic cancer biology and genetics. Nature Reviews Cancer 2(12), 897–909 (2002)
4. Brezniceanu, M.L., Volp, K., Bosser, S., Solbach, C., Lichter, P., et al.: HMGB1 inhibits cell death in yeast and mammalian cells and is abundantly expressed in human breast carcinoma. FASEB Journal 17, 1295–1297 (2003)
5. Cascinu, S., Scartozzi, M., et al.: COX-2 and NF-kB overexpression is common in pancreatic cancer but does not predict for COX-2 inhibitors activity in combination with gemcitabine and oxaliplatin. American Journal of Clinical Oncology 30(5), 526–530 (2007)
6. Ciliberto, A., Novak, B., Tyson, J.: Steady states and oscillations in the p53/Mdm2 network. Cell Cycle 4(3), 488–493 (2005)
7. Clarke, E.M., Emerson, E.A., Sifakis, J.: Model checking: algorithmic verification and debugging. Commun. ACM 52(11), 74–84 (2009)
8. Clarke, E.M., Grumberg, O., Peled, D.A.: Model Checking. MIT Press (1999)
9. Downward, J.: Targeting RAS signalling pathways in cancer therapy. Nature Reviews Cancer 3, 11–22 (2003)
10. Dumitriu, I.E., Baruah, P., Valentinis, B., et al.: Release of high mobility group box 1 by dendritic cells controls T cell activation via the receptor for advanced glycation end products. The Journal of Immunology 174, 7506–7515 (2005)
11. Eddy, S.F., Guo, S., et al.: Inducible IkB kinase/IkB kinase expression is induced by CK2 and promotes aberrant Nuclear Factor-kB activation in breast cancer cells. Cancer Research 65, 11375–11383 (2005)
12. Ellerman, J.E., Brown, C.K., de Vera, M., Zeh, H.J., Billiar, T., et al.: Masquerader: high mobility group box-1 and cancer. Clinical Cancer Research 13, 2836–2848 (2007)
13. Faeder, J.R., Blinov, M.L., Hlavacek, W.S.: Rule-based modeling of biochemical systems with BioNetGen. Methods in Molecular Biology 500, 113–167 (2009)
14. Geva-Zatorsky, N., Rosenfeld, N., Itzkovitz, S., Milo, R., Sigal, A., Dekel, E., Yarnitzky, T., Liron, Y., Polak, P., Lahav, G., Alon, U.: Oscillations and variability in the p53 system. Molecular Systems Biology, 2:2006.0033 (2006)

15. Gillespie, D.T.: A general method for numerically simulating the stochastic time evolution of coupled chemical reactions. Journal of Computational Physics 22(4), 403–434 (1976)
16. Gong, H., Guo, Y., Linstedt, A., Schwartz, R.: Discrete, continuous, and stochastic models of protein sorting in the Golgi apparatus. Physical Review E 81(1), 011914 (2010)
17. Gong, H., Sengupta, H., Linstedt, A., Schwartz, R.: Simulated de novo assembly of Golgi compartments by selective cargo capture during vesicle budding and targeted vesicle fusion. Biophysical Journal 95, 1674–1688 (2008)
18. Gong, H., Zuliani, P., Komuravelli, A., Faeder, J.R., Clarke, E.M.: Analysis and verification of the HMGB1 signaling pathway. BMC Bioinformatics (2010) (to appear)
19. Hanahan, D., Weinberg, R.A.: The hallmarks of cancer. Cell 100(1), 57–70 (2000)
20. Hinz, M., Krappmann, D., Eichten, A., Heder, A., Scheidereit, C., Strauss, M.: NF-κB function in growth control: regulation of cyclin D1 expression and G0/G1-to-S-phase transition. Mol. Cell Biol. 19, 2690–2698 (1999)
21. Hlavacek, W.S., Faeder, J.R., Blinov, M.L., Posner, R.G., Hucka, M., Fontana, W.: Rules for modeling signal-transduction system. Science STKE 2006 re6 (2006)
22. Hoffmann, A., Levchenko, A., Scott, M.L., Baltimore, D.: The IκB-NFκB signaling module: Temporal control and selective gene activation. Science 298, 1241–1245 (2002)
23. Huang, Z.: Bcl-2 family proteins as targets for anticancer drug design. Oncogene 19, 6627–6631 (2000)
24. Idel, S., Dansky, H.M., Breslow, J.L.: A20, a regulator of NFκB, maps to an atherosclerosis locus and differs between parental sensitive C57BL/6J and resistant FVB/N strains. Proceedings of the National Academy of Sciences 100, 14235–14240 (2003)
25. Jha, S.K., Clarke, E.M., Langmead, C.J., Legay, A., Platzer, A., Zuliani, P.: A Bayesian Approach to Model Checking Biological Systems. In: Degano, P., Gorrieri, R. (eds.) CMSB 2009. LNCS, vol. 5688, pp. 218–234. Springer, Heidelberg (2009)
26. Kang, R., Tang, D., Schapiro, N.E., Livesey, K.M., Farkas, A., Loughran, P., Bierhaus, A., Lotze, M.T., Zeh, H.J.: The receptor for advanced glycation end products (RAGE) sustains autophagy and limits apoptosis, promoting pancreatic tumor cell survival. Cell Death and Differentiation 17(4), 666–676 (2009)
27. Krishna, S., Jensen, M.H., Sneppen, K.: Minimal model of spiky oscillations in NF-kB signaling. Proceedings of the National Academy of Sciences 103, 10840–10845 (2006)
28. Langmead, C.J.: Generalized queries and bayesian statistical model checking in dynamic bayesian networks: Application to personalized medicine. In: CSB, pp. 201–212 (2009)
29. Larris, S., Levine, A.J.: The p53 pathway: positive and negative feedback loops. Oncogene 24, 2899–2908 (2005)
30. Lee, D.F., Huang, M.C.: Advances in targeting IKK and IKK-related kinases for cancer therapy. Clinical Cancer Research 14, 5656 (2008)
31. Lipniacki, T., Hat, T., Faeder, J.R., Hlavacek, W.S.: Stochastic effects and bistability in T cell receptor signaling. Journal of Theoretical Biology 254, 110–122 (2008)
32. Lipniacki, T., Paszek, P., Brasier, A., Luxon, B., Kimmel, M.: Crosstalk between p53 and nuclear factor-kB systems: pro-and anti-apoptotic functions of NF-kB. Journal of Theoretical Biology 228, 195–215 (2004)
33. Lotze, M.T., Tracey, K.: High-mobility group box 1 protein (HMGB1): nuclear weapon in the immune arsenal. Nature Reviews Immunology 5, 331–342 (2005)

34. McInnes, C.: Progress in the evaluation of CDK inhibitors as anti-tumor agents. Drug Discovery Today 13(19-20), 875–881 (2008)
35. Mengel, B., Krishna, S., Jensen, M.H., Trusina, A.: Theoretical analyses predict A20 regulates period of NF-κB oscillation. arXiv: bio-ph 0911.0529 (2009)
36. Nelson, D.E., Ihekwaba, A.E.C., et al.: Oscillations in NF-κB signaling control the dynamics of gene expression. Science 306, 704–708 (2004)
37. Nevins, J.R.: The Rb/E2F pathway and cancer. Human Molecular Genetics 10, 699–703 (2001)
38. Puszynski, K., Hat, B., Lipniacki, T.: Oscillations and bistability in the stochastic model of p53 regulation. Journal of Theoretical Biology 254, 452–465 (2008)
39. Rizk, A., Batt, G., Fages, F., Soliman, S.: On a Continuous Degree of Satisfaction of Temporal Logic Formulae with Applications to Systems Biology. In: Heiner, M., Uhrmacher, A.M. (eds.) CMSB 2008. LNCS (LNBI), vol. 5307, pp. 251–268. Springer, Heidelberg (2008)
40. Rotblat, B., Ehrlich, M., Haklai, R., Kloog, Y.: The Ras inhibitor farnesylthiosalicylic acid (salirasib) disrupts the spatiotemporal localization of active Ras: a potential treatment for cancer. Methods in Enzymology 439, 467–489 (2008)
41. Semino, C., Angelini, G., Poggi, A., Rubartelli, A.: NK/iDC interaction results in IL-18 secretion by DCs at the synaptic cleft followed by NK cell activation and release of the DC maturation factor HMGB1. Blood 106, 609–616 (2005)
42. Sherr, C.J., McCormick, F.: The Rb and p53 pathways in cancer. Cancer Cell 2, 103–112 (2002)
43. Tang, X., Liu, D., Shishodia, S., Ozburn, N., Behrens, C., Lee, J.J., Hong, W.K., Aggarwal, B.B., Wistuba, I.I.: Nuclear factor-κB (NF-κB) is frequently expressed in lung cancer and preneoplastic lesions. Cancer 107, 2637–2646 (2006)
44. Vakkila, J., Lotze, M.T.: Inflammation and necrosis promote tumour growth. Nature Reviews Immunology 4, 641–648 (2004)
45. van Beijnum, J.R., Buurman, W.A., Griffioen, A.W.: Convergence and amplification of toll-like receptor (TLR) and receptor for advanced glycation end products (RAGE) signaling pathways via high mobility group B1. Angiogenesis 11, 91–99 (2008)
46. Vogelstein, B., Lane, D., Levine, A.J.: Surfing the p53 network. Nature 408, 307–310 (2000)
47. Wee, K.B., Aguda, B.D.: Akt versus p53 in a network of oncogenes and tumor suppressor genes regulating cell survival and death. Biophysical Journal 91, 857–865 (2006)
48. Wu, H., Lozano, G.: NF-κB activation of p53. a potential mechanism for suppressing cell growth in response to stress. J. Biol. Chem. 269, 20067–20074 (1994)
49. Yao, G., Lee, T.J., Mori, S., Nevins, J., You, L.: A bistable Rb-E2F switch underlies the restriction point. Nature Cell Biology 10, 476–482 (2008)
50. Younes, H.L.S., Simmons, R.G.: Statistical probabilistic model checking with a focus on time-bounded properties. Information and Computation 204(9), 1368–1409 (2006)
51. Zuliani, P., Platzer, A., Clarke, E.M.: Bayesian statistical model checking with application to simulink/stateflow verification. In: HSCC, pp. 243–252 (2010)

Composability: Perspectives
in Ecological Modeling

Ozan Kahramanoğulları[1], Ferenc Jordán[1], and Corrado Priami[1,2]

[1] The Microsoft Research – University of Trento,
Centre for Computational and Systems Biology
[2] Department of Information Engineering and Computer Science,
University of Trento

Abstract. The multiplicity of ecological interactions acting in parallel calls for novel computational approaches in modeling ecosystem dynamics. Composability, a key property of process algebra-based models can help to manage complexity and offer scalable solutions in ecological modeling. We discuss and illustrate how composability of process algebra language constructs can be used as a language aid in the construction of complicated ecosystem models.

Keywords: ecology, modeling, stochastic process algebra, BlenX.

1   Introduction

The systems approach to biology [Kitano, 2002] is now broadly established. Catalyzed by the advances in computational capabilities and the introduction of promising technologies from various disciplines, formal modeling and analysis methodologies are now becoming one of the common instrument-ensembles in biological research. The contribution of the systems point of view to the experimental biology is twofold. Firstly, formal models enforce a rigorous representation of the biological knowledge. This results in disambiguous descriptions of the mechanistic behavior of the biological systems under study. Secondly, simulation and analysis with formal models often provide insights into implicit aspects of the biological systems, and deliver predictions that help biologists to design further experiments.

The algorithmic approach to systems biology [Priami, 2009], driven by the application of core computer science technologies, is based on describing the capabilities of the components of biological systems and their interactions in terms of discrete state spaces. The topological structure and quantitative aspects of algorithmic models mediate various simulation and analysis techniques that are adapted from computer science, and shed light to the mechanistic understanding of the biological systems they model. For example, stochastic simulations provide a means to observe the emergent behavior of the modeled systems, while static analysis capabilities, such as reachability queries on the state space, help to address topological properties.

K. Horimoto, M. Nakatsu, and N. Popov (Eds.): ANB 2011, LNCS 6479, pp. 136–148, 2012.

One of the underlying metaphors of algorithmic systems biology is the perception of biological systems as complex, reactive, information processing systems, where system components interact with each other in diverse ways, and generate new patterns of interaction. Such a consideration makes concurrency theory an appropriate formal framework for studying biological systems with respect to the parallel, distributed and mobile interactions exhibited by these systems. In this regard, the field of *process algebra* provides the principles for defining specific programming primitives and draws the guidelines for designing algorithms that are tailored for biology [Regev and Shapiro, 2002]. In particular, *composability* of the algebraic operators, which is commonly exploited in modeling of computer systems with process algebras, becomes also instrumental while building biological models. This is because composability makes it possible to specify the meaning of a system component in terms of its components and the meaning of the algebraic operator that composes them. As a result of this, each component of a biological system can be modeled independently, allowing large models to be constructed by composition of simple components. Moreover, because one can work on individual components, modifications to the dynamics of a model can be made locally on the appropriate component of the model without modifying the rest of the model.

Although the pioneering efforts in systems biology can be attributed to the differential equation models of Lotka and Volterra of the predator-prey interactions in fisheries [Lotka, 1927,Volterra, 1926], the systems approach is rather underrepresented in ecology in comparison to molecular biology (see [Ulanowicz, 1986,Platt et al., 1981] for early discussions on ecological processes). This can be partly due to diverging considerations of ecosystems, on one hand, in terms of general principles and universal laws, and on the other hand, in terms of phenomena that emerge as a result of vastly parallel, stochastic processes, driven by local rules. This latter perspective, which is closer to the algorithmic approach to systems biology, is emphasized in ecology within *individual based models* (IBM) [DeAngelis and Gross, 1992,Grimm, 1999,Grimm et al., 2006] or *agent based* models (ABM). Another discussion in ecological modeling with parallels to both IBMs and algorithmic systems biology is based on the consideration of ecosystems as complex adaptive systems in which patterns at higher levels emerge from localized interactions and selection processes acting at lower levels [Levin, 1998].

IBMs build on the observations above by emphasizing the ideas that individuals of an ecosystem are different and the interactions between individuals take place locally. Based on these assumptions, IBMs describe populations of systems in terms of discrete and autonomous individuals with distinguished properties. Models built this way are then studied by tracking their individuals, also in terms of their collective behavior through space and time. The aim here is to understand the implications of the local interactions to the whole system with respect to the emerging patterns of behavior during stochastic simulations, and this way link mechanisms to behaviors [Seth, 2007].

The ideal scenario in IBMs is that a model with as little detail as possible reveals as much as possible during simulation. As in molecular biology models, in IBMs there is often a trade-off between simpler, more abstract models and models that reflect more aspects of reality. The main challenge here is to summarize the knowledge on nature accurately into a model, also by resorting to an appropriate level of abstraction. The model should capture the key aspects of each individual's capabilities in terms of its interactions with others and the environment while remaining simple enough for a fruitful analysis. However, IBMs pose this challenge with an additional twist: while the complexity of cellular processes comes mostly from how many (and how many kinds of) molecules interact, an important component of ecological complexity is how many ways components can interact with each other. In ecosystems, several types of interactions act in parallel and they are also in interaction with each other (e.g. [Billick and Case, 1994]). Understanding and modeling the interactions of interactions, as well as finding appropriate common currencies for their quantification are among the most important motives in community and systems ecology [Vasas and Jordán, 2006].

Having emerged as an area of computer science, process algebras profit from a theoretical foundation that provides a rich arsenal of formal techniques and tools as well as a broadly expanding culture of software engineering. In the following, we argue that stochastic process algebra languages may contribute to ecology models [Priami and Quaglia, 2004,Priami, 2009], partly resolving the challenges that confront IBMs. As an evidence for this, we present process algebra representations of ecosystem models and primitives, where composability is the essential ingredient for extending and refining models at different levels, and for designing specialized modeling interfaces. For the models, we use the stochastic process algebra language BlenX [Dematté et al., 2008,Dematté et al., 2010].

2   The BlenX Language

In this section, we provide a brief introduction to stochastic process algebras, in particular, the BlenX language.

Process algebras are formal languages, which were originally introduced as a means to study the properties of complex reactive systems. In these systems, concurrency, that is, the view of systems in which potentially interacting computational processes are executing in parallel, is a central aspect. Due to their capability to capture such a form of concurrency, the process algebra languages qualify as appropriate tools for describing the dynamics of biological systems [Regev and Shapiro, 2002].

BlenX [Dematté et al., 2008,Dematté et al., 2010] is a stochastic process algebra language that shares features with stochastic pi-calculus [Priami, 1995] and Beta-binders [Degano et al., 2005]. As these other members of the family of process algebra-based languages, BlenX has a strong focus on the interactions of entities. BlenX is explicitly designed to model biological entities and their interactions. It is a stochastic language in the sense that the probability and speed of the interactions and actions are specified in the programs that are written in this language. In this respect, BlenX provides an efficient implementation of the Gillespie algorithm [Gillespie, 1977], the semantics of which is given by continuous time Markov-chains. BlenX is a part of the software platform *CoSBiLab*.

In BlenX, each individual is given with an abstract entity that we call a *box*. Each box has a number of connectivity interfaces called *binders*, and it is equipped with an internal program. The sites of interaction are represented as binders on the box surface. For example, in Figure 1, each box has only one binder. Binders are identified by their names, e.g., x and their types, e.g., X.

The mechanism, realized by the interfaces and the internal program govern the interactions of the box and their effects on the box: this mechanism describes a number of possible actions with which the individual can evolve to a new state possibly by interacting with others or on its own. At each simulation step, the simulation engine picks an action of the model in a manner which is biased by the rates of the actions with respect to the Gillespie algorithm. This gives rise to a model behavior in the form of a sequence of model actions that can be read as a time series, depicting the behavior of the model components.

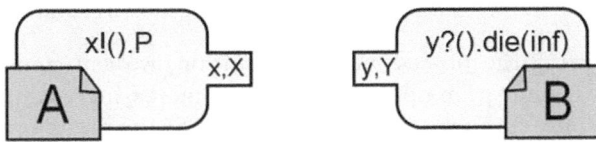

Fig. 1. Two BlenX boxes representing two interacting species A and B

A BlenX model consists of two parts, where the first part contains a description of all the boxes of the model, together with their binders. The second part of the model contains a list of compatibilities of different binders with respect to their types. With respect to the compatibilities described in this part, binders can bind or unbind to binders of other boxes, or perform communications with them to exchange information. For example, with respect to the model given in Figure 1, the compatibility expression $(X, Y, 0, 0, 1)$ indicates that the binders with types X and Y can communicate with a rate 1. The third and forth parameters of this expression state the binding and unbinding rates between these types, which are 0 in this case.

A box can stochastically interact with another box, and change state as a result of this interaction with respect to the actions specified in its internal program. Alternatively, a box can autonomously change state by stochastically performing an action that is given in its internal program. For instance, the interaction of a predator A and its prey B can be described in a BlenX model with the boxes depicted in Figure 1. The interaction rate, specified in the BlenX code, determines the rate of the predation being modeled. The internal program, which can be **nil**, describes this interaction and its consequences in terms of the *actions* the box can undertake. The `nil` action does nothing. Other stochastic *actions* that a BlenX box can perform are summarized as follows: a box can

(*i.*) communicate with another box (or with itself) by performing an input action, e.g., `x?(message)` that is complementary to the output action, e.g., `x!(message)`, of the other box, or vice versa, and this way send or receive a message;

(*ii.*) perform a stochastic `delay` action;

(*iii.*) change (`ch`) the type of one of its interfaces;

(*iv.*) eliminate itself by performing a `die` action;

(*v.*) `expose` a new binder;

(*vi.*) `hide` one of its binders;

(*vii.*) `unhide` a binder which is hidden.

In addition to these actions, there are also other programming constructs available such as if-then statements and state-checks. For example, let us consider the box A in Figure 1. We can define the program P such that it changes the type X to Z if this box is bound to another speices via its interface x:

```
if (x,X) and (x,bound) then ch(x,Z) endif
```

In BlenX, following the process algebra tradition, we can compose actions by using algebraic composition operators to define increasingly complex behaviors. We can sequentially compose actions by resorting to the prefix-operator, which is written as an infix dot. For instance, `ch(x,Z).hide(x).nil` denotes a program that first performs change action and then hides the changed binder. Programs can be composed in parallel. Parallel composition (denoted by the infix operator |, for instance `P|Q`) allows the description of programs, which may run independently in parallel and also synchronize on *complementary actions* (i.e., *input* and *output* over the same channel). Programs can also be composed by *stochastic choice*, denoted with the summation operator "+". The sum of processes `P` and `Q`, `P + Q` behaves either as `P` or as `Q`, determined by their stochastic rates, and selection of one discards the other forever.

In BlenX, we use *events*, which are programming constructs for expressing actions that are enabled by global conditions. For example, in ecosystem models, we use the `new` construct to introduce new individuals of a species to the system, for instance, to model migration or birth, or to implement global influences on the model individuals such as change of seasons.

3   Composability as a Modeling Aid

As we illustrate in the previous section, in BlenX, model expressions are written by resorting to the algebraic notion of composition. In a model, the capabilities of an individual is described by composing atomic BlenX actions within the internal program of the box that models the individual. This way, we can define each model individual in terms of its potential behavior with respect to its interactions

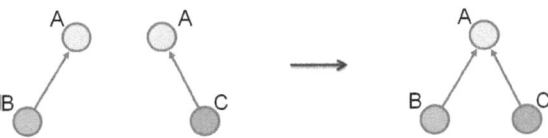

Fig. 2. Composition of two predator-prey interactions, providing an apparent competition module [Holt, 1977]. In the two simple models, A preys on either B or C. In the composed model, A preys on both B and C

and state changes as an algebraic expression. The meaning of this expression is thus delivered in terms of the meaning of its action components and the composition operators, giving rise to composability.

Composability of model expressions becomes an instrumental aid in modeling complex interactions and dynamics such as those that can be observed among individuals of ecosystems. From an analytic bottom-up point of view, composability of BlenX language allows the modelers to consider parts of a complex ecological system in isolation, and build models by composing these parts at the same level. This makes it easy to build larger models when required.

As for illustration for bottom-up composability, let us consider two simple predator-prey models in isolation as depicted in Figure 2. In the first model, species A exclusively preys only on species B. In the second model, species A preys on species C. In BlenX, we can consider these two models in isolation, and simply merge the two models in order to obtain a composed model as depicted in Figure 2 and Figure 3. Then in the composed model, the only additional requirement is the inclusion of the compatibility parameters, which contain the information on which boxes interact with each other.

From a dual top-down point of view, composability makes it possible to consider a component of a model as a black-box, or "open the black-box" to modify the model to include further aspects. This kind of composability becomes instrumental, for example, when a model is refined with a previously ignored detail of the modeled system. Here what we mean by refinement concerns the structure

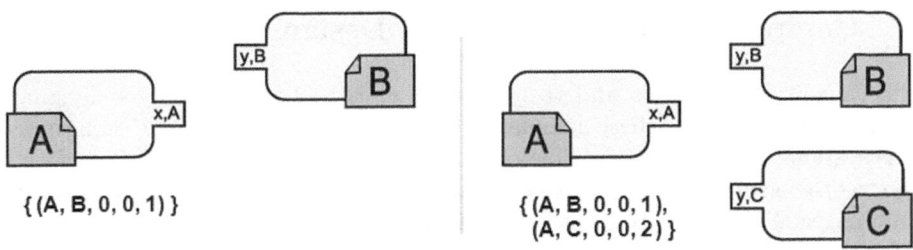

Fig. 3. Graphical representation of the composition in BlenX of the models depicted in Figure 2

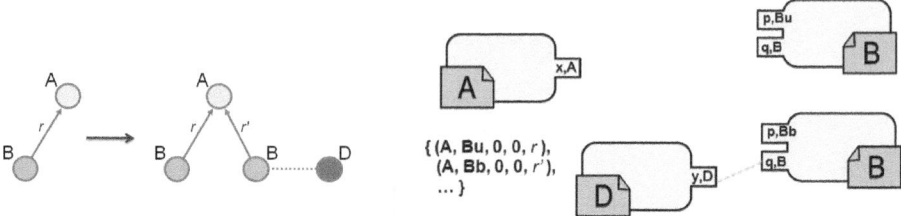

Fig. 4. Refinement of a simple model by further aspects of the predator-prey interaction. In the simple model A preys on B with rate r. In the refined model, A preys on B with rate r' if B is accompanied by D

of the model (rather than size), which boils down to choosing the appropriate component of the model and extending it. Within BlenX, algebraic constructs aid to refine the model by making it possible to work locally on the boxes representing the involved species and including the new data on the system. For example, consider a model, as depicted in Figure 4, where we have a species A that preys on species B. However, in this case, we refine the model with the information that in the presence of species D, the feeding rate of A on B changes from r to r' [Wootton, 1993].

By providing the means for these two dual perspectives, composability becomes instrumental in extending and refining the models to capture the optimal level of representation with respect to the goals of the model [Grimm et al., 2006]. Refined models reflect more aspects of the reality in the structure of the model with respect to the modeled ecosystem. However, sensitivity analysis and other static and dynamic analysis of the model and the simulations with it become more involved as the level of refinement increases. Moreover, the choice on which aspects of "reality" to include in the model in terms of qualitative and quantitative data is an important decision. In this respect, composability becomes an instrumental modeling aid that provides the means to move between different levels of abstraction and extend, and shrink the model with respect to requirements.

4   Composability and Language Design

The algebraic operators and the language constructs of specialized languages such as BlenX allow these languages to capture the mechanistic structure of the systems that they model. This way, for instance, BlenX can capture with its syntax the ecological phenomena that are otherwise challenging to model in other languages. This is an advantage in contrast to other programming languages that can provide a complex encoding of the desired behavior of these phenomena. However, the mathematical syntax of these languages makes them difficult to use for the people who lack training in these languages. This results in a barrier for these languages to be used effectively by a broader audience.

As a remedy for this barrier, one of the central objectives of algorithmic systems biology is the development of user-friendly interface languages for modeling. These interface languages profit from the expressive power of specialized languages, such as BlenX, while remaining accessible to domain specialists who are not familiar with formal languages. In this respect, when stochastic process algebra languages such as BlenX are considered as target languages for user-friendly interface languages, composability becomes a valuable feature that brings an ease to the design and development process.

The CoSBiLab LIME [Kahramanoğulları et al., 2011] is an example to such interface languages for ecosystem modeling. LIME is a language interface to BlenX for building ecosystem models. LIME allows the user to give a biologically intuitive model description in a narrative style controlled natural language. After performing static analysis on the model structure, the LIME translation software tool translates the model description into a BlenX program. This makes the BlenX and CoSBiLab modeling framework handy and intuitive for ecologists. Simulations can be run by the BetaWB and the output can be analyzed and visualized by both Plotter [Dematté et al., 2008] and CoSBiLab Graph [Valentini and Jordán, 2010].

An example LIME model is depicted in Figure 5. A LIME input file can consist of five parts that describe different aspects of the model.

1. The first part is a single statement on the simulation duration.
2. The second part consists of sentences that describe the interactions of the individuals of the modelled ecosystem. Each sentence describes an interaction in the ecosystem together with the ecological patch where it happens and its rate. The interactions can be of four different kinds: *predator-prey, plant-pollinator, direct competition,* and *facilitation.*
3. The optional third part of the input file collects the information on the birth and death rates of the species. Each sentence in this part describes the birth and/or death rates for each species in each habitat patch. If a habitat patch is not specified, the rate is distributed and applies to all the patches: this way, general rates can be defined.
4. The optional fourth part of the input file contains the information on patch dynamics of the ecosystem: each sentence here describes the migration rate between two particular patches of a given species.
5. The fifth part provides the information on the initial population sizes (at the beginning of the simulation).

For an illustrative example, consider the LIME model in Figure 5 that consists of two habitat patches, X and Y, with identical parameters. The local communities consist of two predators, A and B, sharing a single prey, C. The consumption rate of A and B on C are 0.4 and 0.8, respectively. However, A migrates between the patches, with rate 1. A and B have a death rate of 0.0001, and C has a birth rate of 10. Finally, the initial population size is 5 for the predators and 15 for the prey. The LIME description of this model consists of the frames given in Figure 5, which is much easier to write and work with in comparison to the BlenX code that it generates. In order to see this explicitly, let us consider the

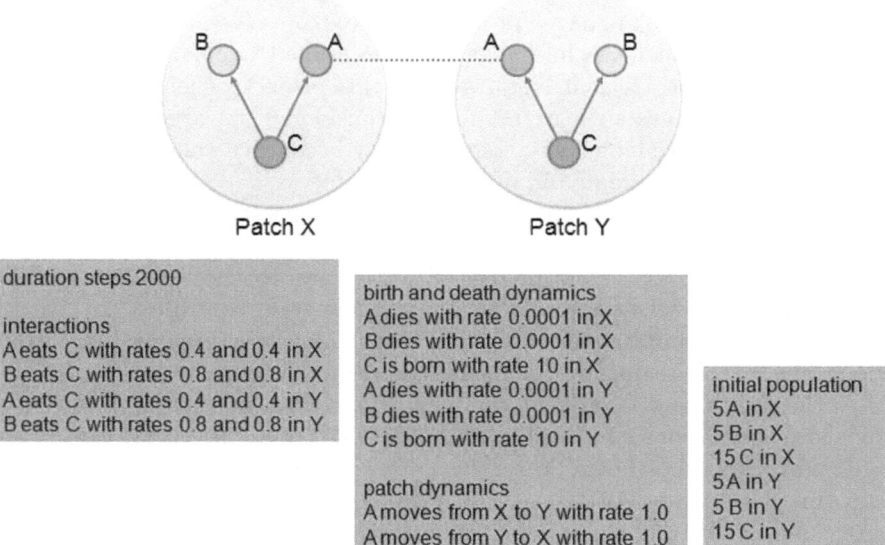

Fig. 5. A graphical representation of an ecosystem model and its CoSBiLab LIME representation. The species A and B are two predators which both prey on C. The species A can move between the patches X and Y

following sentence:

A eats C with rates 0.4 and 0.4 in X

This sentence is translated into BlenX code, a part of which is

```
... if (ex,Aex_X) then ex!().start!() endif
  + if (ey,Aey_X) then ey!().ch(r,ArRep_X) endif ...
```

and the sentence

A dies with rate 0.0001 in X

is translated to the following code:

```
... + if (ex,Aex_X) then die(0.00010) endif ...
```

Migration between patches in the narrative form

A moves from X to Y with rate 1.0

is translated to the following BlenX expressions:

```
... + if (ex,Aex_X) then delay(1.0).ch(r,Ar_Y).
          ch(ex,Aex_Y).ch(ey,Aey_Y).start!() endif ...
```

In comparison to BlenX, LIME has a limited expressive power, since it can be used to model only certain kinds of ecological interactions and dynamics. However, LIME enjoys a higher level of composability in comparison to BlenX. This is because LIME models can be written, extended and modified with a great ease with almost no prior knowledge of this language. For example, the food web models that we borrow from [Pimm, 1980] in Figure 6 can be written within few minutes. This capability makes it very easy to experiment with different models.

Modeling different types of concurrent interspecific interactions is a big challenge in ecology [Olff et al., 2009]. There are many works in ecology literature on single interaction types, with an emphasis on food webs, and an increasing focus on plant-pollinator networks in isolation. However, in ecosystems, different kinds of interactions always happen in parallel, thus it is paramount to model them simultaneously. In LIME, composability of the process algebra constructs play a key role in expressing these different kinds of interactions in a unified manner in a single model. Moreover, composability brings a great ease into the design and maintenance of such interface languages.

5   Discussion

For both technical and historic reasons, individual based models (IBMs) still face major computational challenges [Gronewold and Sonnenschein, 1998]. Some of these challenges are linking elegantly mechanisms and behavior [Seth, 2007], minimizing the combinatorial explosion in state space of complex models, and modeling common currency-based integration of several, multi-level and multi-scale processes [Allen and Starr, 1982,Levin et al., 1997]. In particular, a desired advancement in IBMs is the development of generic mechanisms for integrating multiple, parallel ecological processes [Levin et al., 1997,Olff et al., 2009]. Similar to processes addressed by the language LIME, these processes can act at the same level, for example, as in predation and facilitation interactions among the species of an ecosystem [Bertness and Callaway, 1994], or they can act at different organizational levels, for example, as in dispersal in the metacommunity and competition in the local community. Progress in these problems could help in better understanding the relationship between patterns and processes in ecology [Pimm, 1991].

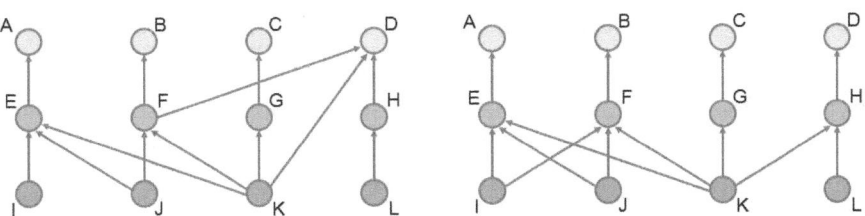

Fig. 6. Graphical representation of two food webs that are variations of each other

The view of ecological systems as complex reactive systems, similar to now broadly established consideration of molecular biology systems, provides the means to model, simulate and analyze these systems, and also indicates directions which can contribute to the solution of above mentioned challenges that confront IBMs. In fact, the consideration of ecosystems as complex reactive systems has parallels also with the complex adaptive system view of of ecosystem models, which are summarized by three properties [Levin, 1998]: (*i.*) diversity and individuality, (*ii.*) qualitative and quantitative aspects of the localized interactions, and (iii:) autonomous processes that reflect the effects of the interactions to their replication or removal, and to the enhancement of their interactions. As we have demonstrated above, composability properties and language constructs of the language BlenX are promising tools for addressing these properties.

The computer simulations in ecology promise the further added value of filling the void due to the impossibility of experiments within the study of certain ecosystems. This is simply because it is not plausible, even if possible, to 'knock out' all the members of a species in a certain ecosystem in order to see the effect. However, certain experiments can be easily designed in silico by resorting to languages such as BlenX and LIME.

An important aspect of the BlenX language with respect to ecological modeling is its stochastic semantics. Stochasticity, which manifests itself as fluctuations in simulations, is an instrumental feature for studying the inherent noise in ecosystems, both at individual level and at the population level. Certain sources of noise are much more important in small populations [Powell and Boland, 2009]. For example, as opposed to deterministic population-level models, stochastic IBMs make it possible to model actual extinction events. While the extinction of rare species is at the frontier of applied ecological research and conservation biology, more generalistic discussions of ecosystem models are limited in both handling the noisy behavior of small populations and modeling extinction. For instance, the same amount of living biomass can behave quite stochastically if it corresponds to two whale individuals, and quite deterministically if it corresponds to millions of organisms in the zooplankton. Availability of stochasticity together with composability is an additional asset from the point of view of IBMs.

References

Allen and Starr, 1982. Allen, T.F.H., Starr, T.B.: Hierarchy: Perspectives for Ecological Complexity. The University of Chicago Press (1982)

Bertness and Callaway, 1994. Bertness, M.D., Callaway, R.: Positive interaction in communities. Trends in Ecology and Evolution 9, 191–193 (1994)

Billick and Case, 1994. Billick, I., Case, T.J.: Higher order interactions in ecological communities: what are they and how can they be detected? Ecology 75, 1529–1543 (1994)

DeAngelis and Gross, 1992. DeAngelis, D.L., Gross, L.J.: Individual-based Models and Approaches in Ecology. Chapman and Hall, New York (1992)

Degano et al., 2005. Degano, P., Prandi, D., Priami, C., Quaglia, P.: Beta binders for biological quantitative experiments. ENTCS 164(3), 101–117 (2005)

Dematté et al., 2010. Dematté, L., Larcher, R., Palmisano, A., Priami, C., Romanel, A.: Programming biology in BlenX. Systems Biology for Signaling Networks 1, 777–821 (2010)

Dematté et al., 2008. Dematté, L., Priami, C., Romanel, A.: The beta workbench: a computational tool to study the dynamics of biological systems. Briefings in Bioinformatics 9, 437–449 (2008)

Dematté et al., 2008. Dematté, L., Priami, C., Romanel, A.: The Blenx Language: A Tutorial. In: Bernardo, M., Degano, P., Tennenholtz, M. (eds.) SFM 2008. LNCS, vol. 5016, pp. 313–365. Springer, Heidelberg (2008)

Gillespie, 1977. Gillespie, D.T.: Exact stochastic simulation of coupled chemical reactions. Journal of Physical Chemistry 81, 2340–2361 (1977)

Grimm, 1999. Grimm, V.: Ten years of individual-based modelling in ecology: what have we learned and what could we learn in the future? Ecological Modelling 115, 129–148 (1999)

Grimm et al., 2006. Grimm, V., Berger, U., Bastiansen, F., Eliassen, S., Ginot, V., Giske, J., Goss-Custard, J., Grand, T., Heinz, S.K., Huse, G., Huth, A., Jepsen, J.U., Jørgensen, C., Mooij, W.M., Müller, B., Peer, G., Piou, C., Railsback, S.F., Robbins, A.M., Robbins, M.M., Rossmanith, E., Rüger, N., Strand, E., Souissim, S., Stillman, R.A., Vabø, R., Visser, U., DeAngelis, D.L.: A standard protocol for describing individual-based and agent-based models. Ecological Modelling 198(1-2), 115–126 (2006)

Gronewold and Sonnenschein, 1998. Gronewold, A., Sonnenschein, M.: Event-based modelling of ecological systems with asynchronous cellular automata. Ecological Modelling 108(1-3), 37–52 (1998)

Holt, 1977. Holt, R.D.: Predation, apparent competition, and the structure of prey communities. Theoretical Population Biology 12(2), 197–229 (1977)

Kahramanoğulları et al., 2011. Kahramanoğulları, O., Jordán, F., Lynch, J.: CoSBi-Lab LIME: A language interface for stochastic dynamical modelling in ecology. Environmental Modelling and Software 26, 685–687 (2011)

Kitano, 2002. Kitano, H.: Systems biology: A brief overview. Science 295, 1662–1664 (2002)

Levin, 1998. Levin, S.A.: Ecosystems and the biosphere as complex adaptive systems. Ecosystems 1(5), 431–436 (1998)

Levin et al., 1997. Levin, S.A., Grenfell, B., Hastings, A., Perelson, A.S.: Mathematical and computational challenges in population biology and ecosystems science. Science 275(5298), 334–343 (1997)

Lotka, 1927. Lotka, A.J.: Fluctuations in the abundance of a species considered mathematically. Nature 119, 12 (1927)

Olff et al., 2009. Olff, H., Alonso, D., Berg, M.P., Eriksson, B.K., Loreau, M., Piersma, T., Rooney, N.: Parallel ecological networks in ecosystems. Philosophical Transactions of Royal Society B 364(1524), 1755–1779 (2009)

Pimm, 1980. Pimm, S.L.: Food web design and the effects of species deletion. Oikos 35, 139–149 (1980)

Pimm, 1991. Pimm, S.L.: The Balance of Nature? Ecological Issues in the Conservation of Species and Communities. The University of Chicago Press (1991)

Platt et al., 1981. Platt, T., Mann, K.H., Ulanowicz, R.E.: Mathematical Models in Biological Oceanography. The UNESCO Press (1981)

Powell and Boland, 2009. Powell, C.R., Boland, R.P.: The effects of stochastic population dynamics on food web structure. Journal of Theoretical Biology 257(1), 170–180 (2009)

Priami, 1995. Priami, C.: Stochastic π-calculus. The Computer Journal 38(6), 578–589 (1995)

Priami, 2009. Priami, C.: Algorithmic systems biology. Communications of the ACM 52(5), 80–89 (2009)

Priami and Quaglia, 2004. Priami, C., Quaglia, P.: Modelling the dynamics of biosystems. Briefings in Bioinformatics 5(3), 259–269 (2004)

Regev and Shapiro, 2002. Regev, A., Shapiro, E.: Cellular abstractions: Cells as computation. Nature 419, 343 (2002)

Seth, 2007. Seth, A.K.: The ecology of action selection: insights from artificial life. Philosophical Transactions of Royal Society B 362(1485), 1545–1558 (2007)

Ulanowicz, 1986. Ulanowicz, R.E.: Growth and Development: Ecosystems Phenomenology. Springer, Heidelberg (1986)

Valentini and Jordán, 2010. Valentini, R., Jordán, F.: CoSBiLab Graph: the network analysis module of CoSBiLab. Environmental Modelling and Software 25, 886–888 (2010)

Vasas and Jordán, 2006. Vasas, V., Jordán, F.: Topological keystone species in ecological interaction networks: considering link quality and non-trophic effects. Ecological Modelling 196(3-4), 365–378 (2006)

Volterra, 1926. Volterra, V.: Fluctuations in the abundance of species considered mathematically. Nature 118, 558–560 (1926)

Wootton, 1993. Wootton, J.T.: Indirect effects and habitat use in an intertidal community: interaction chains and interaction modifications. The American Naturalist 141(1), 71–89 (1993)

A General Procedure for Accurate Parameter Estimation in Dynamic Systems Using New Estimation Errors

Masahiko Nakatsui[1], Alexandre Sedoglavic[2], François Lemaire[2],
François Boulier[2], Asli Ürgüplü[2], and Katsuihisa Horimoto[1,*]

[1] Computational Biology Research Center, National Institute of Advanced Industrial Science and Technology, 2-4-7 Aomi, Koto-ku, Tokyo, Japan
[2] Lille Computer Science Laboratory, University of Science and Technology of Lille, 59655 Villeneuve d'Ascq Cédex, France

Abstract. The investigation of network dynamics is a major issue in systems and synthetic biology. One of the essential steps in a dynamics investigation is the parameter estimation in the model that expresses biological phenomena. Indeed, various techniques for parameter optimization have been devised and implemented in both free and commercial software. While the computational time for parameter estimation has been greatly reduced, due to improvements in calculation algorithms and the advent of high performance computers, the accuracy of parameter estimation has not been addressed.

We previously proposed an approach for accurate parameter optimization by using Differential Elimination, which is an algebraic approach for rewriting a system of differential equations into another equivalent system. The equivalent system has the same solution as the original system, and it includes high-order derivatives, which contain information about the form of the observed time-series data. The introduction of an equivalent system into the numerical parameter optimizing procedure resulted in the drastic improvement of the estimation accuracy, since our approach evaluates the difference of not only the values but also the forms between the measured and estimated data, while the classical numerical approach evaluates only the value difference. In this report, we describe the detailed procedure of our approach for accurate parameter estimation in dynamic systems. The ability of our approach is illustrated in terms of the parameter estimation accuracy, in comparison with classical methods.

1 Introduction

The investigation of network dynamics is a major issue in systems and synthetic biology[1]. In general, a network model for describing the kinetics of constituent molecules is first constructed with reference to the biological knowledge, and then the model is mathematically expressed by differential equations, based on

---

* Corresponding author.

K. Horimoto, M. Nakatsu, and N. Popov (Eds.): ANB 2011, LNCS 6479, pp. 149–166, 2012.

the chemical reactions underlying the kinetics. Finally, the kinetic parameters in the model are estimated by various parameter optimization techniques[2], from the time-series data measured for the constituent molecules. While the computational time for parameter estimation has been greatly reduced, due to the improvements in calculation algorithms and the advent of high performance computers, the accurate numerical estimation of parameter values for a given model remains a limiting step. Indeed, the parameter values estimated by various optimization techniques are frequently quite variable, due to the conditions for parameter estimation, such as the initial values. In particular, we cannot always obtain the data measured for all of the constituent molecules, due to limitations of measurement techniques and ethical constraints. In this case, one of the issues we should resolve is that the parameters are estimated from the data for only some of the constituent molecules. Unfortunately, it is quite difficult to estimate the parameters in such a network model including unmeasured variables.

Differential elimination was applied[3] to improve the parameter estimation methods, especially in the model dynamics including unmonitored variables. The idea consisted of computing differential equations from the input system, from which the unmonitored variables were eliminated. These differential equations could then be used to guess the initial values for the Newton-type numerical parameter optimization scheme. The overall method was implemented over the BLAD libraries[4]. Differential elimination theory is a branch of the differential algebra of Ritt and Kolchin[5], [6]. Its basis was developed by Ritt, who founded the theory of characteristic sets. Ritt's ideas were subsequently developed by Seidenberg [7], Wu[8], Boulier et al.[9],[10] and many other researchers. The Rosenfeld-Gröbner algorithm[9], [10] is the first complete algorithm for differential elimination ever implemented. It relies on Ritt and Seidenberg's ideas, on the Rosenfeld Lemma, which reduces differential problems to non-differential polynomial ones, and on the Gröbner bases theory for solving non-differential polynomial systems (although recent implementations completely avoid Gröbner bases computations). The Rosenfeld-Gröbner algorithm was implemented in 1996 in the diffalg package of the MAPLE computer algebra software. Starting from MAPLE 14, it should be replaced by the MAPLE DifferentialAlgebra package, which relies on the BLAD libraries[11].

Recently, we proposed a new procedure for optimizing the parameters, by using differential elimination. Our procedure partially utilizes a technique from a previous study[12], [13], regarding the introduction of differential elimination into parameter optimization in a network. Instead of using differential elimination for estimating the initial values for the following parameter optimization, the equations derived by differential elimination are directly introduced as the constraints into the objective function for the parameter optimization[14], [15], [16], [17]. Here, we will describe the detailed procedure of our approach, by using a simple model represented as non-linear differential equations. We also discuss the merits and pitfalls of our procedure, in terms of its extension to more realistic and complex models.

2 Procedure

2.1 Overview of Present Procedure

The key point of this study is the introduction of new constraints obtained by differential elimination into the objective function, to improve the parameter accuracy. This section outlines our new procedure for estimating the parameters, using constraints built from differential elimination, and compared it with the classical constraints based on the total relative error. For clarity, the method is described using an academic example.

We first present the example. We then show how to build our new constraints using differential elimination, and how to optimize the evaluation of those new constraints over numeric values. Subsequently, we present our genetic algorithm for estimating the parameter values, and finish with the results. All Maple commands used for computing the expressions described in the following subsections are provided in appendix A.

2.2 Example

Differential algebra aims at studying differential equations from a purely algebraic point of view[5], [6]. Differential elimination theory is a sub theory of differential algebra, based on Rosenfeld-Gröbner[9]. Differential elimination rewrites the inputted system of differential equations to another equivalent system, according to (order of terms). Here, we provide an example of differential elimination, as shown below, according to Boulier[12].

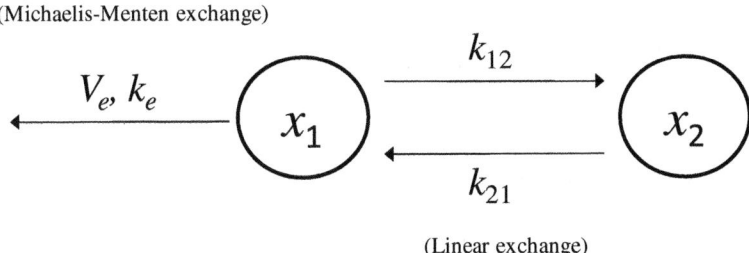

(Michaelis-Menten exchange)

V_e, k_e x_1 k_{12} x_2 k_{21}

(Linear exchange)

Fig. 1. Schematic representation of the model
The model is composed of two state variables, x_1 and x_2. We assumed that the time-series data for one of the variable, x_1, are obtained.

Assume a model of two variables, x_1 and x_2, as schematically depicted in Fig. 1, with the corresponding system of differential equations expressed as follows:

$$\begin{cases} \dot{x}_1 = -k_{12}x_1 + k_{21}x_2 - \frac{V_e x_1}{k_e + x_1} \\ \dot{x}_2 = k_{12}x_1 - k_{21}x_2 \end{cases} \tag{1}$$

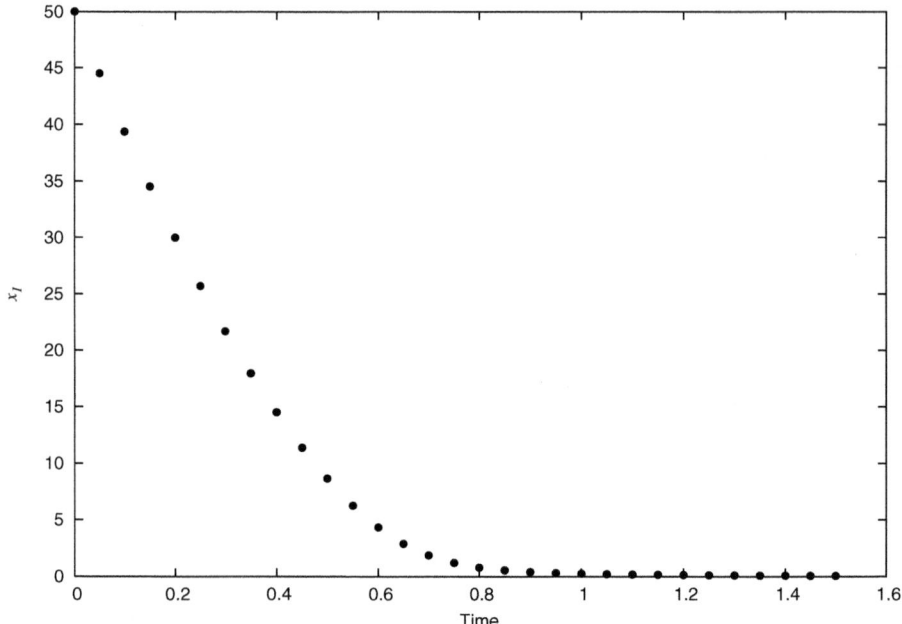

Fig. 2. Reference curve

According to the kinetics of the model (Eqn. (1)), a reference curve of one variable, x_1, was generated for $0 \leq t \leq 1.5$ with intervals of 0.05, under the following conditions: $x_1(0) = 50.0, x_2(0) = 0.0, Ve = 101.0, k_{12} = 0.5, k_{21} = 3.0$ and $k_e = 7.0$.

where k_{12}, k_{21}, k_e and V_e are some constants. Two molecules are assumed to bind according to Michaelis-Menten kinetics.

Here we assume that the time-series of only one variable, x_1, can be observed. x_2 is assumed to be non-observed; however, we assumed that $x_2(0) = 0$ was known. According to the model in Fig. 1, a reference curve of one variable, x_1, was generated in Fig. 2. Among the parameters in the model, the values of three parameters, k_{12}, k_{21}, and V_e, were estimated, and the values of the remaining parameters were set to the same values as those used in the generation of the reference curve of Fig. 2.

2.3   Differential Elimination

The differential elimination then produces the following two equations equivalent to the above system.

$$\begin{cases} \dot{x}_1(k_{21} + x_1) + k_{21}x_1^2 + (k_{12} + V_e)x_1 - k_{21}(k_e + x_1)x_2 = 0 \\ \ddot{x}_1(x_1 + k_e)^2 + (k_{12} + k_{21})\dot{x}_1(x_1 + k_e)^2 + V_e\dot{x}_1k_e + k_{21}V_ex_1(x_1 + k_e) = 0 \end{cases} \tag{2}$$

As a consequence, the latter two equations should be zero for any solution of (1). The latter to equations, respectfully, called $C_{1,t}$ and $C_{2,t}$ in the following, will be used to define our error estimation, based on the evaluation of $|C_{1,t}| + |C_{2,t}|$.

System (2) can be computed in Maple 14, using the following commands:

```
> with(DifferentialAlgebra):
> sys := [
>     x1[t] - ( -k12*x1 + k21*x2 - Ve*x1/(ke+x1)),
>     x2[t] - ( k12*x1 - k21*x2)
>     ];
                            Ve x1
     sys := [x1[t] + k12 x1 - k21 x2 + -------, x2[t] - k12 x1 + k21 x2]
                            ke + x1

> R := DifferentialRing(blocks=[x2,x1,k12(),k21(),Ve(),ke()], derivations=[t]);
                      R := differential_ring

> Ids := RosenfeldGroebner( numer(sys), denom(sys), R,
>                           basefield=field(generators=[k12,k21,Ve,ke]));
                      Ids := [regular_differential_chain]

> eqs := Equations(Ids[1]);
eqs := [
                                                          2
    k21 x2 x1 + k21 x2 ke - x1[t] x1 - x1[t] ke - k12 x1  - k12 x1 ke - Ve x1,

           2                                2        2
    x1[t, t] x1  + 2 x1[t, t] x1 ke + x1[t, t] ke  + x1[t] x1  k12

          2                                                          2
    + x1[t] x1  k21 + 2 x1[t] x1 k12 ke + 2 x1[t] x1 k21 ke + x1[t] k12 ke

              2                   2          2
    + x1[t] k21 ke  + x1[t] Ve ke + x1  k21 Ve + x1 k21 Ve ke]
```

2.4 Simplification

In general, the problem of reducing the evaluation complexity (additions, multiplications) is difficult and requires a large number of computer operations (a.k.a. a high algorithmic complexity). Moreover, the evaluation complexity of the Rosenfeld-Gröbner output tends to be exponential in the evaluation complexity of the input, especially when using elimination rankings, as in this case. Consequently, before directly applying techniques such as factorization, Horner schemes, common sub expression detection, etc. for reducing the evaluation complexity, we try to use the knowledge we already have on the initial ODE system.

We now describe a preprocessing step that facilitates the evaluation of $\tilde{C}_{DE} = |C_{1,t}| + |C_{2,t}|$.

The expressions of $C_{1,t}$ and $C_{2,t}$ given in (2) are not the expressions originally computed by the Rosenfeld-Gröbner algorithm. Indeed, the Rosenfeld-Gröbner algorithm outputs expanded expressions.

Thus, using the Rosenfeld-Gröbner outputs, one has to evaluate the following expression, $\tilde{C}_{DE}$:

$$\bar{C}_{DE} = \left| -k_{21}x_2 k_e - k_{21}x_2 x_1 + \dot{x}_1 k_e + \dot{x}_1 x_1 + k_{12}x_1 k_e + k_{12}x_1^2 + V_e x_1 \right| \qquad (3)$$
$$+ \left| k_{21} k_e V_e x_1 + 2k_e k_{12}x_1 \dot{x}_1 + 2k_{21} k_e \dot{x}_1 x_1 + 2k_e \ddot{x}_1 x_1 + k_{12}\dot{x}_1 k_e^2 \right.$$
$$\left. + k_e V_e \dot{x}_1 + k_{12}x_1^2 \dot{x}_1 + k_{21} k_e^2 \dot{x}_1 + k_{21}x_1^2 \dot{x}_1 + k_{21}x_1^2 V_e + \ddot{x}_1 k_e^2 + \ddot{x}_1 x_1^2 \right|$$

requiring 18 additions + 46 multiplications (+2 function evaluations for the absolute value). These operations represent the evaluation complexity of the expression $\bar{C}_{DE}$.

Since the expressions of $C_{1,t}$ and $C_{2,t}$ were computed from an ODE system involving the denominator $k_e + x_1$, from a Michaelis-Menten factor, the expression $k_e + x_1$ can be likely be factorized. By introducing a new variable, $d_e = k_e + x_1$, and applying the substitution $k_e \to d_e - x_1$ in the previous expression of $\bar{C}_{DE}$, one gets

$$\bar{C}_{DE} = \left| -k_{21}x_2 d_e + \dot{x}_1 d_e + k_{12}x_1 d_e + V_e x_1 \right| \qquad (4)$$
$$+ \left| k_{21} V_e x_1 d_e + k_{12}\dot{x}_1 d_e^2 + V_e \dot{x}_1 d_e - V_e \dot{x}_1 x_1 + k_{21}\dot{x}_1 d_e^2 + \ddot{x}_1 d_e^2 \right|$$

requiring 9 additions + 21 multiplications.

Please note that the last expression of $\bar{C}_{DE}$ does not involve k_e anymore, which shows that the variable k_e only appears in $\bar{C}_{DE}$ in the term $k_e + x_1$.

This trick with the denominators has divided the number of operations by 2. On more complex systems, the benefit can be much greater. It is worth noting that the trick works quite similarly if several denominators are involved and if each denominator linearly involves a parameter that is not involved in the other denominators. More precisely, if one has n denominators of the form $k_i + f_i$, and if k_i is not involved in any f_i, then one performs n substitutions $k_i \to f_i - d_i$.

Further computations using a Horner scheme can now be accomplished. For example, applying a recursive Horner scheme with decreasing priority on the variables $d_e, x_1, x_2, \dot{x}_1, \ddot{x}_1$ yields:

$$\bar{C}_{DE} = \left| V_e x_1 - (k_{21}x_2 - \dot{x}_1 - k_{12}x_1)d_e \right| \qquad (5)$$
$$+ \left| -V_e \dot{x}_1 x_1 + (k_{21} V_e x_1 + V_e \dot{x}_1 + (\ddot{x}_1 + (k_{12} + k_{21})\dot{x}_1)d_e)d_e \right|$$

requiring 9 additions + 12 multiplications.

To finish, further simplification can be achieved using the optimize command of the optimize package in the Computer Algebra software Maple. This last command tries to recognize common expressions in order to compute common subexpressions only once. This command is not very costly, since it is based on easy heuristics. In our case, it yields the sequence of commands:

$$t7 = \left| V_e x_1 - (k_{21}x_2 - \dot{x}_1 - k_{12}x_1)d_e \right|, \qquad (6)$$
$$t8 = V_e \dot{x}_1,$$
$$t19 = \left| -t8 x_1 + (k_{21} V_e x_1 + t8 + (\ddot{x}_1 + (k_{12} + k_{21})\dot{x}_1)d_e)d_e \right|,$$
$$C_{DE} = t7 + t19$$

requring 9 additions + 11 multiplications + 4 assignments. Note that the last gain here is only 1 multiplication, but can be higher on larger systems.

All previous operations can be automated in Maple (see appendix A for the complete set of Maple commands); the C command of the optimize package yields the C code as

```
t7  =  fabs (Ve*x1−(k21*x2−x1t−k12*x1)*de);
t8  =  Ve*x1t;
t19 =  fabs(−t8*x1+(k21*Ve*x1+t8+(x1tt+(k12+k21)*x1t)*de)*de);
E   =  t7+t19;.
```

2.5   Introduction of Constraints

The objective function for parameter optimization in this study is composed of two terms: one is the standard error function between the estimated and monitored data, and the other is the constraints obtained by differential elimination. The error function is defined as follows: Suppose $x^c_{i,t}$ is the time-series data at time t of x_i, simulated by using the estimated parameter values and the model equations by integration, and $x^m_{i,t}$ represents the monitored data at time t. The sum of the absolute values of the relative error between $x^c_{i,t}$ and $x^m_{i,t}$ gives the averaged relative error over the numbers of monitored variables and time points, E, as a standard error function, i.e.,

$$E = \frac{1}{NT} \sum_{i=1}^{N} \sum_{t=1}^{T} \left| \frac{x^c_{i,t} - x^m_{i,t}}{x^m_{i,t}} \right| \tag{7}$$

where N and T are the numbers of monitored variables and time points, respectively.

Next we define the DE constraints obtained by the differential elimination and simplification procedure. The simplified equivalent system (Eqn. (6)) is composed of x_1, its derivatives ($\dot{x}_1$ and $\ddot{x}_1$), x_2, and the parameters (k_{12}, k_{21}, V_e and k_e). Note that x_2 in Eqn. (6) can be estimated by x_1, the parameters, and $x_2(0)$. The derivatives of variable x_1 can be estimated numerically by the following procedure. First, we obtain two equations by a Taylor expansion of $x_1(t)$,

$$x_1(t + h) = x_1(t) + hx'_1(t) + \frac{h^2}{2}x''_1(t) + \frac{h^3}{6}x'''_1(t) + \cdots, \tag{8}$$

$$x_1(t - h) = x_1(t) - hx'_1(t) + \frac{h^2}{2}x''_1(t) - \frac{h^3}{6}x'''_1(t) + \cdots. \tag{9}$$

Second, we subtract Eqn. (9) from (8),

$$x_1(t + h) - x_1(t - h) = 2hx'_1(t) + \frac{1}{3}h^3 x'''_1(t) + \cdots, \tag{10}$$

$$2hx'_1(t) = x_1(t + h) - x_1(t - h) - \frac{1}{3}h^3 x'''_1(t) + \cdots,$$

$$x'_1(t) = \frac{x_1(t + h) - x_1(t - h)}{2h} - \frac{h^2}{6}x'''_1(t) + \cdots.$$

Finally, we obtain following approximation, under the assumption of $0 < h < 1$,

$$x_1'(t) = \frac{x_1(t + h) - x_1(t - h)}{2h} + O(h^2). \tag{11}$$

We are able to obtain higher-order derivatives from lower-order derivatives in same way, as mentioned above. For instance, we can estimate second order derivatives of x_1 by using following equation,

$$x_1''(t) = \frac{x_1'(t + h) - x_1'(t - h)}{2h} + O(h^2). \tag{12}$$

The value of the simplified equivalent system (Eqn. (6)) can be calculated by the substitution of the observed x_1, its numerically the estimated derivatives, estimated x_2, and the parameter values estimated by the numerical parameter optimizing procedure. In general, Differential Elimination rewrites the original system of differential equations into an equivalent system, which means both systems have the same solutions. This clearly shows that the evaluated values of the equivalent system will be zero with exactly estimated parameter sets, time-series data without noise, and derivatives. Thus, the equivalent system can be regarded as a kind of objective function that expresses the difference between the monitored and estimated data. In this study, we express DE Constraint (C_{DE}), as the average of the linear combination of the equation in the equivalent system over the number of equations and time points, as follows:

$$C_{DE} = \frac{1}{LT} \sum_{l=1}^{L} \sum_{t=1}^{T} |C_{l,t}| \tag{13}$$

where L and T are the numbers of equivalent equations and time points, respectively. Finally, we introduce C_{DE} $\bar{C}_{DE}$, which is simplified as C_{DE}, into the objective function, F, in combination with E, as:

$$F = \alpha E + (1 - \alpha)C_{DE} \tag{14}$$

where $\alpha (0 \leq \alpha \leq 1)$ is the weight of the two functions. As a result, our computational task is to find a set of parameter values that minimize F. When we apply the simplification procedure (see 2.4), then $\frac{1}{LT}\bar{C}_{DE}$ is used instead of C_{DE}.

The weighting factor α in the objective function F is estimated from the slope of the Pareto-optimal solutions. First, we obtained some parameter sets (in the case study, we obtained 200 kinds of parameter sets) by the compute_parameter_set function, under the conditions of $\delta = 1.0$ and the tentative value of α ta $= 1$ (this means we used the classical objective function, i.e. $F = E$). Second, we selected the Pareto-optimal solutions from the list of estimated parameter sets, by the select_pareto_optimal_solutions function. By fitting the linear function $C = aE + b$ to the selected the Pareto-optimal solutions, we obtained the slope of Pareto-optimal solutions, a. Finally, we estimated the value of α from the slope a.

The detailed algorithms for estimating the value of α are shown in Algorithm 1 and 2. Fig. 3 represents a part of the estimated parameter sets in the case study (the detailed algorithms for the parameter optimization we used for the case study are shown in 2.6), the Pareto-optimal solutions, and the fitted line for the Pareto-optimal solutions. We obtained the slope $a = 20.7653$ for the case study, and the value of α was estimated as $\alpha = 0.95406$.

2.6   Optimization Algorithm

Our approach can be applied to many kinds of parameter optimizing procedures, such as the Gradient-based method and the evolutionary optimizing method, including the Modified Powell method[18], [19], Genetic Algorithms[20], [21], and Particle Swarm Optimization[22], [23], by modifying the objective function (cost function) only[16].

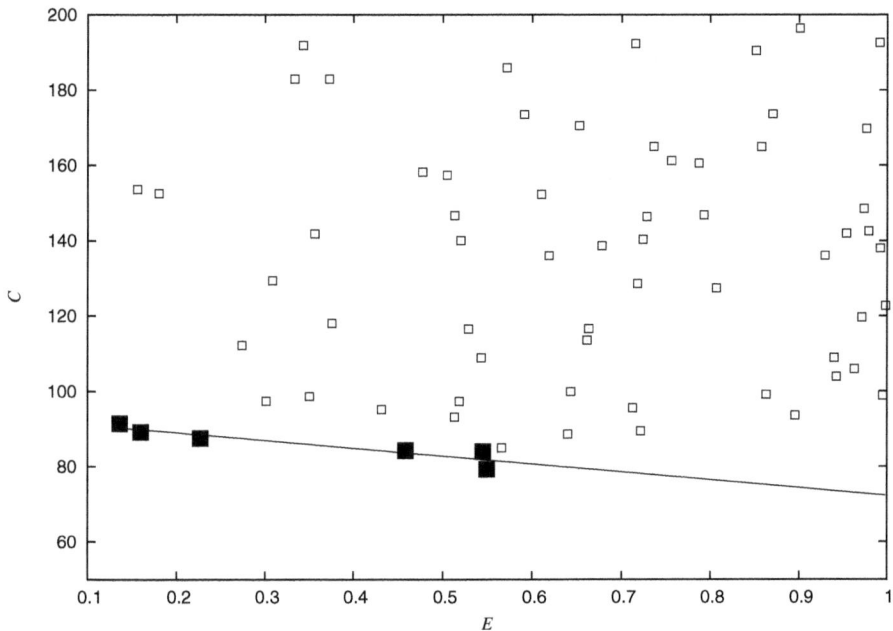

Fig. 3. Slope of Pareto-optimal solutions

The empty squares ($\square$) indicate the set of evaluated values, E and C_{DE}. The filled squares ($\blacksquare$) show the Pareto-optimal solutions, and the line represents the fitted line for Pareto-optimal solutions.

Here, we applied our approach to Real-coded Genetic Algorithms[24], [25], [26], to demonstrate its ability. The detailed algorithms used to analyze the case study (Fig. 1 and 2) are shown in Algorithm 3 to 5.

Algorithm 1. Estimate value of weighting factor α

Function : estimate_alpha(δ, *pop*, *gen*, *trials*)
Input : error tolerance δ, population of GA *pop*, maximum generation counts *gen*, and trial number of GA *trials*
Return : estimated value of weighting factor α

```
1: RES ← compute_parameter_set(α = 1, δ, pop, gen, trials)
2: P ← select_pareto_optimal_solutions(RES)
3: EV ← φ
4: CV ← φ
5: n size of P
6: for i = 0 to n do
7:     EV ← EV union E(Pᵢ)
8:     CV ← CV union C(Pᵢ)
9: end for
10: fit CVᵢ = −aEVᵢ + b from EV and CV by using least square method
11: return a/(a + 1)
```

Algorithm 2. Select Pareto-optimal solutions

Function : select_pareto_optimal_solutions(R)
Input : R set of estimated parameters
Return : Pareto-optimal solutions (P)

```
1: P ← φ
2: EV ← φ
3: CV ← φ
4: n size of R
5: for i = 0 to n do
6:     EV ← EV union E(Rᵢ)
7:     CV ← CV union C(Rᵢ)
8: end for
9: for i = 0 to n do
10:    Flag lp = true
11:    for j = 0 to n do
12:        if !(EVᵢ ≤ EVⱼ and CVᵢ ≤ CVⱼ) then
13:            lp ← false
14:        end if
15:    end for
16:    if lp then
17:        P ← P union Rᵢ
18:    end if
19: end for
20: return P
```

Algorithm 3. Modify the parameter set K by computing the next generation

Function compute_next_generation(α, K)

Input : the weighting factor α, a parameter set K

1: n size of K
2: denote $K = \{k_1, \ldots, k_n\}$
3: compute $1 \leq s \leq n$ such that k_s is the one best element according to the F function (i.e. $F(k_s)$ is the minimum of $F(k_1), \ldots, F(K_n)$)
4: pick a random number r such that $1 \leq r \leq n$, and r is different from s
5: mix k_s and k_r and compute a new set $k' = \{K'_1, \ldots, k'_n\}$
6: $K' \leftarrow K'$ union $\{k_s\}$
7: modify k by replacing k_s and k_r by the two best elements of K' according to the F function

Algorithm 4. Optimization process

Function : compute_one_parameter_set(α, δ, pop, gen)

Input : the weighting factor α, the error tolerance delta for function F, the population size of GA pop, the maximum generation counts gen

Return : a set containing zero or one parameter set

1: create a set K containing pop random parameter sets
2: **for** $i = 1$ to gen **do**
3:     compute_next_generation(α, K)
4:     **if** an element k in K satisfies $E(k) \leq \delta$ **then**
5:         **return** k
6:     **end if**
7: **end for**
8: **return** ϕ

Let us explain the differences between our procedure and the classical constraint E. First of all, by using $\alpha = 1$, one obtains a classical genetic algorithm using the relative error E, since we have $F = E$ when $\alpha = 1$. Second, when using $\alpha < 1$, each parameter set k returned by the compute_parameter_sets satisfies $E(k) \leq \delta$, as in the classical procedure. However, the manner in which the population evolves (in the compute_next_generation) depends on the function F. To summarize, the objective function F is only used to direct the evolution of the population, by not using the objective function F to select the final candidates, and thus it makes sense to compare the parameter sets computed in the classical procedure and in our procedure.

2.7   Results

To evaluate the ability of our procedure, we performed a simulation study by using the objective function with and without the newly developed DE constraints, by estimating the kinetic parameters in Eqn. (1). Here, we assume that

Algorithm 5. Generate a list of estimated parameter sets

Function : compute_parameter_sets(α, δ, pop, gen, $trials$)

Input : the weighting factor α, the error tolerance δ for function F, the population size of GA pop, the maximum generation counts gen, the trial number $trials$

Return : a list of parameter sets

1: $RES \leftarrow \phi$
2: **for** $i = 1$ to $trials$ **do**
3:     $RES \leftarrow RES$ union compute_one_parameter_set(α, δ, pop, gen)
4: **end for**
5: **return** RES

the time-series of only one variable, x_1, can be observed. According to the model, the reference curve of one variable, x_1, was generated in Fig. 2. Among the parameters in the model, the values of three parameters, k_{12}, k_{21}, and V_e, were estimated, and the values of the remaining parameters were set to the same values as those used in the generation of the reference curve.

The introduction of DE constraints into the objective function was quite effective, in the comparison with the distributions of the parameter values estimated with and without DE constraints (see Fig. 4). Indeed, the distribution of the estimated k_{12} and k_{21} values was highly concentrated around the correct values by the estimation with the introduction (Fig. 4 (A)), while the estimated parameters were widely distributed by the estimation without the introduction of DE constraints (Fig. 4 (B)).

3   Discussion

The accuracy of parameter estimation was clearly improved by the introduction of DE constraints into the objective function of the numerical parameter optimizing method. Indeed, the parameter value sets estimated with the introduction of DE constraints into the objective function were sharply distributed near the correct values, in contrast to the wide distribution without the introduction. In general, the derivatives included the information on the curve form of the observed time-series data, such as slope, extremal point and inflection point. This indicates that the new objective function we proposed estimates the difference of not only the values but also the forms between the measured and estimated data, while the classical objective function estimates only the value difference. Note that the DE constraint is rationally reduced from the original system of differential equations for a given model, in a mathematical sense. Thus, our approach is expected to become a general approach for parameter optimization to improve the parameter accuracy.

As expected, the new objective function requires more computational time, in comparison with an objective function with only a standard error function, due to the increase of the function in the DE constraints. In equivalent systems derived by Differential Elimination, the number of terms and operators frequently

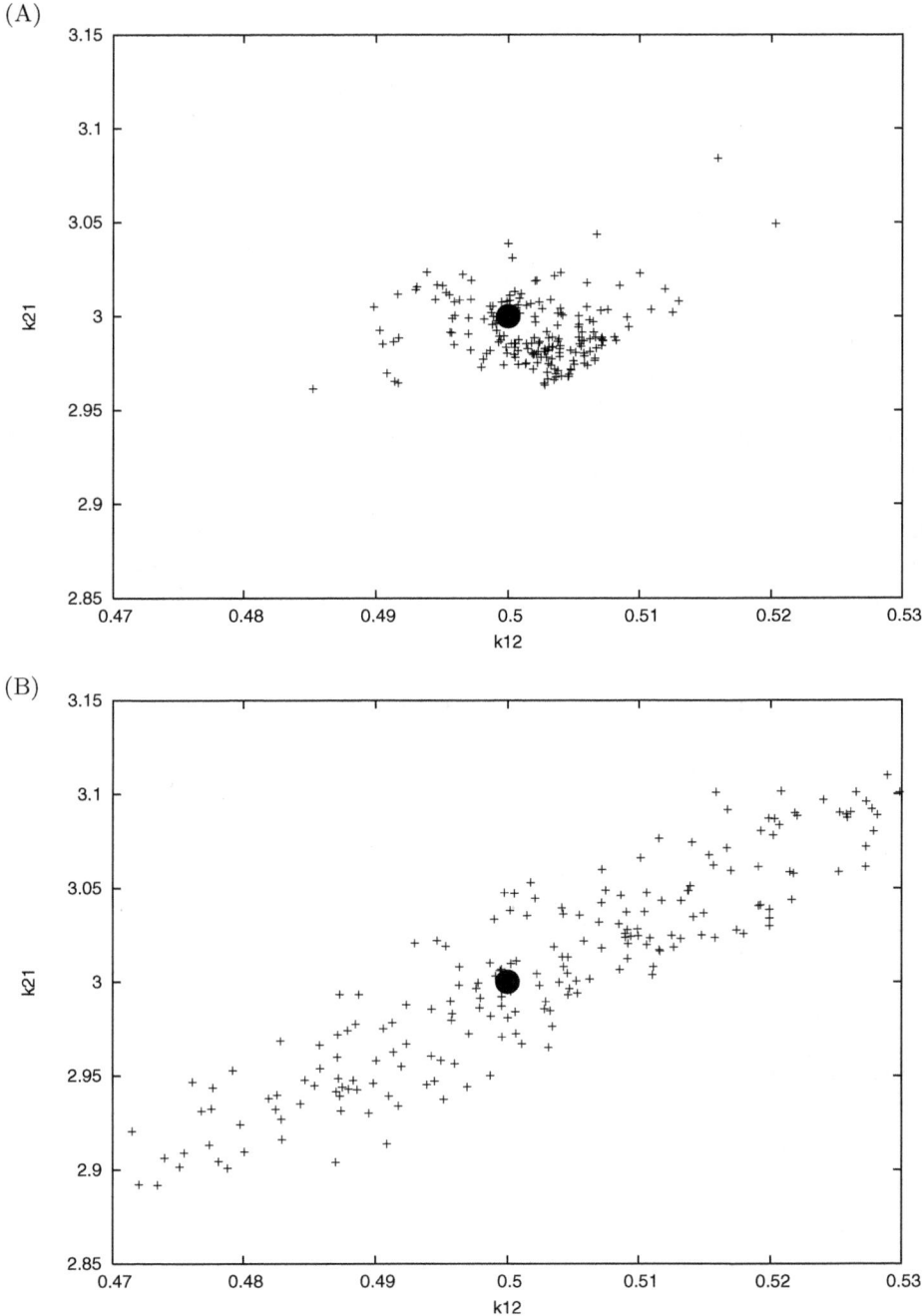

Fig. 4. Comparison of parameter value clouds estimated by the classical or our proposed procedure, (A) with and (B) without DE constraints

The given values are as follows: $x_2(0) = 0.0$ and $k_e = 7.0$. The black circles indicate the correct parameter set.

increases, and this may make the application of our procedure to a complex or large system difficult, without simplification of the equivalent system. To overcome the difficulty in the complex system, we applied simplification by symbolic computation (see 2.4). For instance, we could estimate the kinetic parameters in the negative feed-back oscillator model[27], [28], [29] by using the simplification procedure[17], while the estimation without the simplification failed, due to the immense computational time.

Another possible way to overcome the difficulty in complex models is to approximate the DE constraint. In the DE constraint, the terms with a higher order of derivatives in the differential equations generally appeared in the equivalent system. The magnitudes of the estimated values of the higher order derivatives were relatively smaller than those of the lower order derivatives. Although our procedure was useful, even for noisy data in a simple model[15], the estimated values of the higher order derivatives for noisy data may become large in this case. However, some techniques are frequently used for smoothing noisy data, and after smoothing, the values of the higher order derivatives may be smaller. If the terms with higher order derivatives can be neglected in the estimation, then the computational time may be reduced. Further studies to improve the computational time by approximation of the DE constraint will be reported in the near future.

A   Implementation of Simplification

The following commands use the new DifferentialAlgebra package, and thus require Maple 14 to work.

```
    |\^/|      Maple 14 (X86 64 LINUX)
._|\|   |/|_. Copyright (c) Maplesoft, a division of Waterloo Maple Inc. 2010
 \  MAPLE  /  All rights reserved. Maple is a trademark of
 <____ ____>  Waterloo Maple Inc.
      |       Type ? for help.
  libname := "/home/calforme/lemaire/Triade/src/lib", "/usr/local/maple14/lib"

> with(DifferentialAlgebra):
> with(CodeGeneration):
> with(codegen):
> sys := [
>     x1[t] - ( -k12*x1 + k21*x2 - Ve*x1/(ke+x1)),
>     x2[t] - ( k12*x1 - k21*x2)
>         ];
                              Ve x1
      sys := [x1[t] + k12 x1 - k21 x2 + -------, x2[t] - k12 x1 + k21 x2]
                              ke + x1

>
> R := DifferentialRing(blocks=[x2,x1,k12(),k21(),Ve(),ke()], derivations=[t]);
                      R := differential_ring

> Ids := RosenfeldGroebner( numer(sys), denom(sys), R,
>                      basefield=field(generators=[k12,k21,Ve,ke]));
                  Ids := [regular_differential_chain]
```

```
> eqs := Equations(Ids[1]);
eqs := [

                                            2
   k21 x2 x1 + k21 x2 ke - x1[t] x1 - x1[t] ke - k12 x1  - k12 x1 ke - Ve x1,

           2                           2              2
   x1[t, t] x1  + 2 x1[t, t] x1 ke + x1[t, t] ke  + x1[t] x1  k12

          2                                                                    2
   + x1[t] x1  k21 + 2 x1[t] x1 k12 ke + 2 x1[t] x1 k21 ke + x1[t] k12 ke

          2                  2
   + x1[t] k21 ke  + x1[t] Ve ke + x1  k21 Ve + x1 k21 Ve ke]

# One performs some necessary renaming
> eqs := subs(x1[t,t]=x1tt, x1[t]=x1t, x1[]=x1, x2[t]=x2t, x2[]=x2, eqs);
                                                           2
eqs := [k21 x2 x1 + k21 x2 ke - x1t x1 - x1t ke - k12 x1  - k12 x1 ke - Ve x1,

         2                        2          2              2
   x1tt x1  + 2 x1tt x1 ke + x1tt ke  + x1t x1  k12 + x1t x1  k21

                                            2            2
   + 2 x1t x1 k12 ke + 2 x1t x1 k21 ke + x1t k12 ke  + x1t k21 ke

          2
   + x1t Ve ke + x1  k21 Ve + x1 k21 Ve ke]

> toTransform := [ result = abs(eqs[1]) + abs(eqs[2]) ];
toTransform := [result =

                                                   2
   | -k21 x2 x1 - k21 x2 ke + x1t x1 + x1t ke + k12 x1  + k12 x1 ke + Ve x1 |

          2                        2          2              2
   + | x1tt x1  + 2 x1tt x1 ke + x1tt ke  + x1t x1  k12 + x1t x1  k21

                                            2            2
   + 2 x1t x1 k12 ke + 2 x1t x1 k21 ke + x1t k12 ke  + x1t k21 ke

          2
   + x1t Ve ke + x1  k21 Ve + x1 k21 Ve ke |]

> cost(toTransform);
        18 additions + 2 functions + 46 multiplications + assignments

# One guesses that the denominator ke+x1 appears in many places.
# To make it appear, one introduces de = ke + x1
# and performs the substitution ke -> de - x1
> toTransform2 := subs(ke = de - x1, toTransform):
> toTransform2 := simplify(toTransform2);
toTransform2 := [result = | -k21 x2 de + x1t de + k12 x1 de + Ve x1 | + |

         2            2            2
   x1tt de  + x1t k12 de  + x1t k21 de  + x1t Ve de - x1t Ve x1 + x1 k21 Ve de
```

```
     |]

> cost(toTransform2);
          9 additions + 2 functions + 21 multiplications + assignments

> eqs2 := subs(ke = de - x1, eqs):
> eqs2 := simplify(eqs2);
eqs2 := [k21 x2 de - x1t de - k12 x1 de - Ve x1,

              2            2              2
     x1tt de  + x1t k12 de  + x1t k21 de  + x1t Ve de - x1t Ve x1 + x1 k21 Ve de

     ]

# One remarks that ke does not appear anymore.
# Using horner and optimization.
> eqs3 := convert(eqs2, horner, [de,x1,x2,x1t,x1tt]);
eqs3 := [-Ve x1 + (k21 x2 - x1t - k12 x1) de,

    -x1t Ve x1 + (x1 k21 Ve + x1t Ve + (x1tt + (k12 + k21) x1t) de) de]

> toTransform := [ result = abs(eqs3[1]) + abs(eqs3[2]) ];
toTransform := [result = | Ve x1 - (k21 x2 - x1t - k12 x1) de |

    + | -x1t Ve x1 + (x1 k21 Ve + x1t Ve + (x1tt + (k12 + k21) x1t) de) de |]

> cost( toTransform);
          9 additions + 2 functions + 12 multiplications + assignments

> out := optimize( toTransform );
out := t7 = | Ve x1 - (k21 x2 - x1t - k12 x1) de |, t8 = x1t Ve,

    t19 = | -t8 x1 + (x1 k21 Ve + t8 + (x1tt + (k12 + k21) x1t) de) de |,

    result = t7 + t19

> cost([out]);
          2 functions + 11 multiplications + 9 additions + 4 assignments

# One generates the C code
> C( [ out ]);
      t7 = fabs(Ve*x1-(k21*x2-x1t-k12*x1)*de);
      t8 = x1t*Ve;
      t19 = fabs(-t8*x1+(x1*k21*Ve+t8+(x1tt+(k12+k21)*x1t)*de)*de);
      result = t7+t19;
> quit
memory used=32.7MB, alloc=28.4MB, time=0.18
```

Acknowledgments. This work was partially supported by a project grant, 'Development of Analysis Technology for Induced Pluripotent Stem (iPS) Cell', from The New Energy and Industrial Technology Development Organization (NEDO).

References

1. Kitano, H.: System Biology: A Brief Overview. Science 295(5560), 1662–1664 (2002)
2. Nocedal, J., Wright, S.J.: Numerical Optimization. Springer, New York (1999)
3. Denis-Vidal, L., Joly-Blanchard, G., Noiret, C.: System identifiability (symbolic computation) and parameter estimation (numerical computation). Numerical Algorithms 34, 282–292 (2003)
4. Boulier, F., Denis-Vidal, F., Henin, T., Lemaire, F.: LÉPISME. In: Proceedings of the ICPSS Conference (2004), http://hal.archives-ouvertes.fr/hal-00140368
5. Ritt, J.F.: Differential Algebra. Dover Publications Inc., New York (1950)
6. Kolchin, E.R.: Differential Algebra and Algebraic Groups. Academic Press, New York (1973)
7. Seidenberg, A.: An elimination theory for differential algebra. Univ. California Publ. Math. (New Series) 3, 31–65 (1956)
8. Wu, W.T.: On the foundation of algebraic differential geometry. Mechanization of Mathematics, Research Preprints 3, 2–27 (1989)
9. Boulier, F., Lazard, D., Ollivier, F., Petitot, M.: Representation for the radical of a finitely generated differential ideal. In: Proceedings of ISSAC 1995, pp. 158–166 (1995)
10. Boulier, F., Lazard, D., Ollivier, F., Petitot, M.: Computing representations for radicals of finitely generated differential ideals. Journal of AAECC 20(1), 73–121 (2009); (1997 Techrep. IT306 of the LIFL)
11. Boulier, F.: The BLAD libraries (2004), http://www.lifl.fr/~boulier/BLAD
12. Boulier, F.: Differential Elimination and Biological Modeling. Johann Radon Institute for Computational and Applied Mathematics (RICAM) Book Series 2, 111–139 (2007)
13. Boulier, F., Lemaire, F.: Differential Algebra and System Modeling in Cellular Biology. In: Horimoto, K., Regensburger, G., Rosenkranz, M., Yoshida, H. (eds.) AB 2008. LNCS, vol. 5147, pp. 22–39. Springer, Heidelberg (2008)
14. Nakatsui, M., Horimoto, K.: Parameter Optimization in the network dynamics including unmeasured variables by the symbolic-numeric approach. In: Proc. of the Third International Symposium on Optimization and Systems Biology (OSB 2009), pp. 245–253 (2009)
15. Nakatsui, M., Horimoto, K.: Improvement of Estimation Accuracy in Parameter Optimization by Symbolic Computation. In: Proceedings of IEEE Multi-Conference on Systems and Control (in press)
16. Nakatsui, M., Horimoto, K., Okamoto, M., Tokumoto, Y., Miyake, J.: Parameter Optimization by Using Differential Elimination: a General Approach for Introducing Constraints into Objective Functions. BMC Systems Biology (in press)
17. Nakatsui, M., Horimoto, K., Lemaire, F., Ürgüplü, A., Sedoglavic, F., Boulier, F.: Brute force meets Bruno force in parameter optimization: Introduction of novel constraints for parameter accuracy improvement by symbolic computation. IET Systems Biology (in press)
18. Powell, M.J.D.: An efficient method for finding the minimum of a function of several variables without calculating derivatives. Computer Journal 7, 142–162 (1954)
19. Powell, M.J.D.: On the calculation of orthogonal vectors. Computer Journal 11, 302–304 (1968)
20. Holland, J.H.: Adaptation in Natural and Artificial Systems. The University of Michigan Press, Ann Arbor (1975)

21. Goldberg, D.D.: Genetic Algorithms in Search, Optimization and Machine Learning. Addison-Wesley Longman Publishing Co., Inc., Boston (1989)
22. Eberhart, R., Kennedy, J.: A New Optimizer Using Particle Swarm Theory. In: Proc. of Sixth International Symposium on Micro Machine and Human Science (Nagoya Japan), pp. 39–43. IEEE Service Center, Piscataway (1995)
23. Kennedy, J., Eberhart, R.: Particle swarm optimization. In: Proc. IEEE International Conference on Neural Networks (Perth, Australia), pp. IV:1942– IV:1948. IEEE Service Center, Piscataway (1995)
24. Jonikow, C.Z., Michalewicz, Z.: An Experimental Comparison of Binary and Floating Point Representations in Genetic Algorithms. In: Proceedings of the Fourth International Conference on Genetic Algorithms, pp. 31–36 (1991)
25. Ono, I., Kobayashi, S.: A real-coded genetic algorithm for function optimization using unimodal distribution crossover. In: Proc 7th ICGA, pp. 249–253 (1997)
26. Satoh, H., Ono, I., Kobayashi, S.: A new generation alternation model of genetic algorithm and its assessment. J. of Japanese Society for Artificial Intelligence 15(2), 743–744 (1997)
27. Novák, B., Tyson, J.J.: Design principles of biochemical oscillators. Nat. Rev. Mol. Cell Biol. 9(12), 981–991 (2008)
28. Kwon, Y.K., Cho, K.H.: Quantitative analysis of robustness and fragility in biological networks based on feedback dynamics. Bioinformatics 24(7), 987–994 (2008)
29. Tyson, J.J., Chen, K.C., Novák, B.: Sniffers, buzzers, toggles and blinkers: dynamics of regulatory and signaling pathways in the cell. Curr. Opin. Cell. Biol. 15(2), 221–231 (2003)

Analyzing Pathways Using ASP-Based Approaches

Oliver Ray[1], Takehide Soh[2], and Katsumi Inoue[3]

[1] University of Bristol
Merchant Venturers Building, Woodland Road
Bristol, BS8 1UB, United Kingdom
oray@cs.bris.ac.uk
[2] Graduate University for Advanced Studies
[3] National Institute of Informatics
2-1-2 Hitotsubashi, Chiyoda-ku
Tokyo, 101-8430, Japan
{soh,ki}@nii.ac.jp

Abstract. This paper contributes to a line of research which aims to combine numerical information with logical inference in order to find the most likely states of a biological system under various (actual or hypothetical) constraints. To this end, we build upon a state-of-the-art approach that employs weighted Boolean constraints to represent and reason about biochemical reaction networks. Our first contribution is to show how this existing method fails to deal satisfactorily with networks that contain cycles. Our second contribution is to define a new method which correctly handles such cases by exploiting the formalism of Answer Set Programming (ASP). We demonstrate the significance of our results on two case-studies previously studied in the literature.

1 Introduction

This paper is concerned with an area of research known as Symbolic Systems Biology (SSB) which involves the application of formal methods to biological networks. The main focus of the work is on combining numerical weights with logical inference in order to find the most likely states of a biological system under various (actual or hypothetical) constraints. To this end, we build upon a state-of-the-art approach proposed by Tiwari et al. in [31] for analysing reaction networks using Boolean weighted Maximal-Satisfiability (MaxSat) — a classical NP-complete task for which surprisingly practical solvers now exist [1,20].

Following Tiwari et al., our aim is to use logical inference in the analysis of biological networks, but to combine it with some form of numerical optimisation to provide a more robust and useful approach. Thus, instead of regarding some states as possible and others as impossible, we want to determine the *degree* to which each state might account for a set of given constraints or preferences (representing partial known or desired properties of the system). This allows us to compute, for example, the most likely state that results from a given set of initial inputs or the most likely state that results in a given set of target outputs.

K. Horimoto, M. Nakatsu, and N. Popov (Eds.): ANB 2011, LNCS 6479, pp. 167–183, 2012.
© Springer-Verlag Berlin Heidelberg 2012

Our paper makes two main contributions to this area. The first contribution is to show how the existing MaxSat approach fails to deal satisfactorily with networks that contain cycles (by incorrectly allowing self-supporting pathways that produce their own inputs at no cost). The second contribution is to define a new method which correctly handles these cases by exploiting the formalism of Answer Set Programming (ASP) — a recent paradigm which is closely related to that of Boolean satisfiability but which is specifically designed to penalise logically unfounded loops [21,13] like those arising in this context.

The key insights of this paper can be related to a distinction between two well-known theoretical concepts from the semantics of logic programming: namely, the difference between so-called *stable models* and *supported models* of a program. Intuitively, supported models allow self-supporting loops while stable models do not. It turns out that the maximal-weight network states computed by the MaxSat approach can be expressed as the minimal-cost supported models of a dual answer set program; but that it is in fact the stable models which represent the true cost of a given state. Since ASP solvers can compute both stable and supported models, they offer a practical implementation of both methods.

The rest of the paper is structured as follows. Section 2 reviews the subject of ASP. Section 3 briefly introduces the field of SSB. Sections 4, 5 and 6 are all concerned with reviewing the MaxSat approach of Tiwari et al. [31]. Section 7 identifies some undesirable properties of the MaxSat method when applied to reaction networks with cycles. It also proposes a suitable correction and looks at its implications with respect to Tiwari et al.'s sporulation and MAPk case studies. Section 8 provides an ASP implementation of our corrected approach. Section 9 presents some related work. Section 10 concludes. For completeness, some technical notes are presented at the end of the paper.

2   Answer Set Programming

Answer Set Programming (ASP) [21,13], which can be seen as a variation of Logic Programming (LP) [22], is quickly becoming established as a prominent knowledge representation and reasoning paradigm in its own right. Even though both LP and ASP are built upon the language of first-order normal clauses, their execution mechanisms are rather complementary. The LP approach, as typified by many commercially available Prolog systems, is a top-down method based on depth-first unfolding of clauses. The ASP approach, as popularised by several freely available stable model solvers, is a bottom-up method based on reduction to propositional satisfiability (SAT) solving. While both approaches have advantages and disadvantages, the semantics and proof procedures of ASP are especially well suited to the purposes of this paper.

Hereafter, we use the standard Boolean connectives: and $\wedge$; or $\vee$; not $\neg$; if $\leftarrow$; if and only if $\leftrightarrow$. We use standard ASP constructs: normal rules a :- $l_1, \ldots, l_k$ which ensure the head atom a is true if the body literals $l_1, \ldots, l_k$ are true; optimisation directives $\#minimize[l_1 = w_1, \ldots, l_k = w_k]$ which ensure the sum of the weights w_i on the true literals l_i is minimised; constraint literals

$m\{l_1,\ldots,l_k\}n$ which are true when at least m and at most n literals $l_1,\ldots,l_k$ are true — and where the bounds assume default values $m = 0$ and $n = k$ if they are omitted; pooled literals $p(\ldots, a\,;\,b,\ldots)$ which stand for the two choices $p(\ldots,a,\ldots)$ and $p(\ldots,b,\ldots)$; conditional literals $l_1 : l_2$ which stand for all instances of l_1 that satisfy l_2; and domain constraints $\#domain\ p(V)$ which restrict the domain of a variable V to terms satisfying a predicate p.

In general, ASP supports the use of both classical and default negation. But here it suffices to consider only the latter negation inherited from normal logic programming. Thus, if a is an atom then $not\ a$ is true if there is no reason to believe a. This operator has long been used as a convenient and powerful way of representing and reasoning about defaults and exceptions. Its meaning is formalised by identifying certain models of an Answer Set Program that satisfy certain key properties. Two of the most important are the so-called supported and stable models which, following [9], can be characterised as follows:

Definition 1. *A supported model M of an Answer Set Program P is a set of ground atoms such that for each atom $a \in M$ there is a ground instance of a rule $r \in P$ whose head atom is a and whose body literals are satisfied by M.*

Definition 2. *A stable model M of an Answer Set Program P is a set of ground atoms for which there is a strict well-founded partial order $<$ such that for each atom $a \in M$ there is a ground instance of a rule $r \in P$ whose head atom is identical to a, and whose body literals are satisfied by M, and whose positive body literals all precede a with respect to $<$.*

These notions are closely related to the so-called Clark Completion [6]. As shown in [23]: a model is stable only if it is a minimal Herbrand model of the Clark Completion; and a model is supported if and only if it is a Herbrand model of the Clark Completion. Hence all stable models are also supported models.

3   Symbolic Systems Biology

A considerable amount of current biological knowledge is represented in the form of interaction networks at various levels of cellular abstraction. Much of this information is publicly available from online data sources that specialise in providing graph-based models of biological activity for many organisms at the metabolic, proteomic and genomic levels. While the specific entities and relations denoted by the nodes and edges in such networks are context dependent, the obvious similarities shared by all such formalisms mean that it is possible to develop general-purpose tools which can be used for analysing, interacting, exploiting and integrating all types of information available in this form.

Biological networks are often mapped on to dynamical systems governed by (ordinary or partial) differential equations. But the difficulty of finding suitably accurate rate parameters has begun to motivate the use of more abstract types of reasoning over discrete state spaces. Logic-based approaches are starting to play an important role in these efforts by providing a highly-expressive and human-understandable formalism for representing and reasoning about such networks. This increasing application of logical methods to biological data has given rise a growing area of research known as Symbolic Systems Biology (SSB).

Several logic-based methods have already been successfully exploited in this context, such as action languages [7,2], active learning [17], causal models [3], formal methods [5], integer programming [30], qualitative reasoning [16], spatio-temporal logics [10], term rewriting [8], abduction [15,25], induction [27,24], and their combination [29,26]. These approaches rest upon the common assumption that, to a first approximation, stable biological states can be determined from the network topology with no need for detailed quantitative simulations. This paper is primarily concerned with the MaxSat method of Tiwari et al. [31] which is distinguished by its ability to combine logic and weights.

The use of weights is motivated by the need for more flexibility when tailoring generic reaction networks to particular biological contexts, the need for more feedback when solving problems with too many or too few solutions, and the need for more robustness when modelling noisy observations and/or uncertain knowledge. On the one hand, weights can be set to make specific constraints more or less important in different settings [31, p.157]. On the other hand, weights can be used to rank possible solutions when there are too many of them — or even to suggest likely causes of failure when there are too few [31, p.167].

4   Reaction Networks

Reaction networks have been proposed as a general framework for representing and reasoning about complex biological interactions from metabolic networks to signalling pathways and genetic interactions. Each network defines a set of basic transformations, called *reactions*, between a set of basic biochemical entities, called *species*. Depending on the biological context, species may denote enzymes or metabolites, molecules or ions, ligands or receptors, or proteins in various states of post-transcriptional modification. But, logically, they are just elements s of a given set S. As formalised in [31, p.158]:

Definition 3. *A species s is a member of a fixed set S.*

A reaction is a biochemical transformation which results in the production of one output set of species, called products, from two input sets of species, called re-actants and modifiers. The difference between these two inputs is that reactants undergo a permanent change (which means they are 'used-up' during a reaction) while modifiers only undergo a temporary change (and so can be 're-used' after a reaction). For example, in a metabolic network, products, reactants, and mod-ifiers would be used to model metabolites, substrates, and enzymes, respectively. As formalised in [31, p.158]:

Definition 4. *A reaction* $r = \langle R, M, P \rangle$ *is a triple with three mutually disjoint sets of species* R, M, *and* P, *called* reactants, modifiers, *and* products, *which are said to be* consumed, required, *and* produced *by* r, *respectively. Reactants and modifiers are also called* inputs. *Products are also called* outputs. *The reactants, modifiers, and products of* r *are denoted* $R(r)$, $M(r)$, *and* $P(r)$, *respectively.*

In this way, a network is a set of reactions and a particular problem instance is obtained by specifying a set of given input species (initial species), a set of required output species (target species), and a set of species which should not be produced (forbidden species). In the simplest case, each species is regarded as being present or absent, and each reaction is regarded as being active or inactive. For convenience the state of each individual reaction r_i is represented by a corresponding Boolean variable b_i which is *true* (resp. *false*) if and only if r_i is active (resp. inactive). As formalised in [31, p.158]:

Definition 5. *A network* $N = \{r_1, \ldots, r_n\}$ *is a set of reactions. With respect to such a network, the sets of reactions in which some species* s *occurs as a reactant, modifier, or product are denoted* $R^{-1}(s)$, $M^{-1}(s)$, *and* $P^{-1}(s)$, *respectively.*

Definition 6. *A network instance* $X = \langle N, I, F, T \rangle$ *is a 4-tuple with a network* N *and three sets of species* I, F, *and* T *called* input species, forbidden species, *and* target species, *respectively. Associated with any network* $N = \{r_1, \ldots, r_n\}$ *is a set of Boolean variables* $B = \{b_1, \ldots, b_n\}$, *called* state variables.

The state of an entire network N is given by a set of reactions $\sigma \subseteq N$ which are active. One of the main tasks relating to such networks is that of finding the so-called *equilibrium states* of a network in which target species are consistently produced from initial species by combinations of reactions, called pathways. This is complicated by the fact that different reactions may compete for the same species, which means that a given network can have many viable pathways in which certain reactions are activated at the expense of their competitors.

Definition 7. *Let* $N = \{r_1, \ldots, r_n\}$ *be a network. Two reactions* r_i *and* r_j *are said to* compete *(with each other) if and only if* $i \neq j$ *and there exists a species* s *which is an input of both reactions and which is consumed by at least one.*

5   Equilibrium States

At the heart of their approach, Tiwari et al. aim to show how equilibrium states can be characterised by a set of Boolean constraints over the variables b_i that represent the activities of the reactions r_i. To avoid introducing further variables that explicitly encode the presence of species, they augment a given network with so-called *dummy reactions*. These do not represent real biochemical transform-ations, as they have no inputs, but they do provide a formal way of ensuring that any (initial or imported) species can be present by activating the corresponding dummy reaction.

The definitions below formalise the constraints and dummy reactions in [31]. Definition 8 introduces the predicate *present4i* used by Tiwari et al. to denote the presence of a species s for a reaction i. Definitions 9 and 10 characterise the equilibrium states (of networks and instances) by a set of four constraints over a network augmented with two types of dummy reaction. One dummy reaction, which we call a *completion* reaction, is used to produce each non-initial species with no other producer. Another dummy reaction, which we call an *initialisation* reaction, is used to produce all of the initial species.

As described in [31, p.158], a species s is said to be *present for* a reaction i (as a reactant) if and only if s is actually present (i.e., it is made as the product of some active reaction) and s is not competitively consumed (i.e., it is not used as a reactant of some other active reaction):

Definition 8. *Let S be a set of species and let $N = \{r_1, \ldots, r_n\}$ be a network. Then, for all species $s \in S$ and for all integers $1 \leq i \leq n$, let:*

$$(*) \quad present4i(s, i) \quad = \quad \bigvee_{r_j \in P^{-1}(s)} b_j \quad \wedge \quad \bigwedge_{r_j \in R^{-1}(s), j \neq i} \neg b_j$$

As described in [31, pp.159-162], equilibrium states are characterised by four constraints stating that (A) a reaction is active if and only if all its inputs are present for that reaction, (B) all dummy reactions are inactive, (C) all target species are present, (D) all forbidden species are absent:

Definition 9. *Let N be a network. Let N' be the network, called the* completion *of N, obtained by adding to N one reaction $\langle \emptyset, \emptyset, \{s\} \rangle$ for each species s which is not produced by any reaction in N. Let σ be any subset of N' and assume that a state variable b_k is true if and only if the corresponding reaction r_k is in σ. Then σ is an equilibrium state of N if and only if the following hold:* [1]

$$(A) \quad \text{for all } r_i \in N \quad : \quad b_i \quad \leftrightarrow \quad \bigwedge_{s \in R(r_i) \cup M(r_i)} present4i(s, i)$$

$$(B) \quad \text{for all } r_i \in N'/N \quad : \quad \neg b_i$$

Definition 10. *Let $X = \langle N, I, F, T \rangle$ be a network instance; and let N^* be the network, called the* initialisation *of N, obtained by adding to N one reaction $\langle \emptyset, \emptyset, I \rangle$. Let σ be any equilibrium state of N^*. Then σ is an equilibrium state of X if and only if the following hold:* [2]

$$(C) \quad \text{for all } s \in T \quad : \quad \bigvee_{r_j \in P^{-1}(s)} b_j$$

$$(D) \quad \text{for all } s \in F \quad : \quad \bigwedge_{r_j \in P^{-1}(s)} \neg b_j$$

6   Maximal Solutions

A key feature of Tiwari et al.'s MaxSat approach is that it actually treats all constraints as *soft* requirements with associated weights. Thus the task they consider is not to satisfy all of the constraints, but rather to maximise the total weight of the satisfied constraints. In this way they introduce (one of) two further constraints (E or F), called *hints*, which can be added to Definition 10 above. These express a preference for turning on or turning off as many reactions as possible. As formalised in [31, p.161]:

(E) for all $r_i \in N$: b_i

(F) for all $r_i \in N$: $\neg b_i$

To assign suitable weights $x_A - x_F$ to the constraints (A-F) above, Tiwari et al. suggest a methodology (tailored for signalling pathways) based on three parameters: the number W of reactions in the (augmented) network; the number k of (non-initial) species with no producers; and if there is a preference *pref* for turning reactions 'on' or 'off'. The weights are specified in Definition 11 below (which uses the fact that setting the weight of a constraint to zero is equivalent to omitting that constraint altogether). As formalised in [31, p.162]:

Definition 11. *Let $X = \langle N, I, F, T \rangle$ be a network instance. Let N^* be the initialisation of N. Let N' be the completion of N^*. Let $W = \|N'\|$. Let $k = \|N'/N^*\|$. Let pref be one of three values 'on', 'off', or 'neither'. Let $x_A \ldots x_F$ be integers, called* weights, *defined as follows:*

- $x_A = W$ % importance of correctly using network reaction
- $x_B \approx W/(k+1)$ % importance of not using completion reaction
- $x_C = W$ % importance of producing a target species
- $x_D \gg W$ % importance of avoiding a forbidden species

- $x_E = \begin{cases} 1 & \text{if pref} = \text{'on'} \\ 0 & \text{otherwise} \end{cases}$ % importance of turning on a reaction
  % (if applicable)

- $x_F = \begin{cases} 1 & \text{if pref} = \text{'off'} \\ 0 & \text{otherwise} \end{cases}$ % importance of turning off a reaction
  % (if applicable)

Now let σ be any subset of N' and let $n_A \ldots n_F$ be the number of constraints of the form $(A \ldots F)$, respectively, that are satisfied by σ. Then the weight of σ, denoted $\mu(\sigma)$ is the defined as follows (and any state σ which maximises this value is called a maximal solution *of X):*

$$\mu(\sigma) \quad = \quad \sum_{I \in \{A,B,C,D,E,F\}} n_I \cdot x_I$$

7   Preferred Solutions

The maximal solutions above have some undesirable properties when applied to networks with cycles. This is illustrated by Example 1 below which shows a network N comprising a reversible pathway with four reactions $\{r_1, r_3, r_5, r_7\}$ in one direction (that take a to b to c to d to e) and four reactions $\{r_8, r_6, r_4, r_2\}$ in the other direction (that take e and f to d to c to b to a and f).

If there is one initial species a and one target species e, then there are two maximal solutions which produce the latter from the former. The first activates one linear pathway with reactions $\{r_1, r_3, r_5, r_7\}$. The second activates two cycles: one comprising reactions $\{r_1, r_2\}$ and the other comprising reactions $\{r_7, r_8\}$. In both cases, one extra dummy reaction r_0 (not shown) is also active.

Here, each arrow in a network diagram depicts a reaction, whose product(s) are each denoted by arrow heads, whose reactants(s) are denoted by solid tails, and whose modifier(s) are denoted by dotted tails. The feedback loop through f is merely there to ensure the critique below applies under any preference for turning reactions on or off; but similar arguments would apply without it.

Example 1. Fix a set of species $S = \{a, b, c, d, e, f\}$ and a reaction network $N = \{r_1, r_2, r_3, r_4, r_5, r_6, r_7, r_8\}$ as shown below. Let $X = \langle N, I, F, T \rangle$ be the network instance with one initial species $I_1 = \{a\}$, no forbidden species $F_1 = \emptyset$, and one target species $T_1 = \{e\}$.

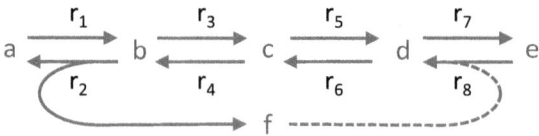

Initialising and completing the network results in the addition of just one dummy reaction $r_0 = \langle \emptyset, \emptyset, \{a\} \rangle$. Hence, $W = 9$, $k = 0$, and so $x_A = x_B = x_C = 9$. It can be shown there are exactly two equilibrium states: $\sigma_1 = \{r_0, r_1, r_3, r_5, r_7\}$ and $\sigma_2 = \{r_0, r_1, r_2, r_7, r_8\}$. It can be shown these two states are the only maximal solutions (under any preference for turning reactions 'off' or 'on').

The first solution is intuitive because all of the active reactions use inputs which can be produced from the initial species. But the second solution is not intuitive because two of the active reactions, r_7 and r_8 use inputs which cannot be produced from the initial species. It could be said that, in effect, σ_2 produces d and e out of 'thin air'.

The problem is that, with no c, neither d nor e can be produced without the other already being present. And, in many applications (as when interpreting this network as a signalling pathway or metabolic network, for example), it would simply not be reasonable to give this self-supporting loop the same weight as the linear pathway from a to e.

This suggests that an alternative way of computing the weight of a state is needed that pays some penalty for violating at least one reaction in every such 'unfounded' loop. This can be done by observing that when there are no problematic loops it is always possible to sequentially order the active reactions in a way which ensures all inputs are the product of some preceding reaction.

Conversely, the non-existence of an ordering with this property indicates that the state has a cyclic pathway that essentially produces its own inputs. But, as formalised in Definition 12 below, finding the minimum number of reactions which violate this property (under all possible orderings) determines how many reactions must be penalised.

Definition 12. *Let N be a network. Let σ be any subset of N and let $\mu(\sigma)$ be the weight of σ (as formalised in Definition 11). Now let $\tau(\sigma)$ be the set of all strict total orders on σ and, for each relation $<$ in $\tau(\sigma)$, let $\epsilon(<)$ be the number of reactions r in σ having an input species s that is not produced by any earlier reaction $r' < r$. Then the* corrected weight $\gamma(\sigma)$ *of σ is defined as follows (and any state σ which maximises this value is called a* preferred solution *of X):* [3]

$$\gamma(\sigma) \quad = \quad \mu(\sigma) \quad - \quad x_A \cdot \min_{< \, \in \, \tau(\sigma)} \epsilon(<)$$

Example 2. Recall the network $N = \{r_1, r_2, r_3, r_4, r_5, r_6, r_7, r_8\}$ from Example 1 above. Let $X' = \langle N, I', F, T \rangle$ be the network instance with no initial species $I' = \emptyset$, no forbidden species $F_1 = \emptyset$, and one target species $T_1 = \{e\}$.

Now there are no dummy reactions so $W = 8$, $k = 0$ and $x_A = x_B = x_C = 8$. This time, there is just one equilibrium state $\sigma_3 = \{r_1, r_2, r_7, r_8\}$ which is the unique maximal solution (under any preference). But, under any ordering of the reactions, at least two will have inputs not produced by an earlier one. Thus the true weight is obtained by subtracting 2*8=16 from the original value.

By contrast, there are six preferred solutions: $\pi_1 = \emptyset$ which violates one constraint (C) for e; $\pi_2 = \{r_7\}$ which violates one constraint (A) for r_7; $\pi_3 = \{r_7, r_8\}$ which violates one constraint (A) for r_8; $\pi_4 = \{r_5, r_7\}$ which violates one constraint (A) for r_5; $\pi_5 = \{r_3, r_5, r_7\}$ which violates one constraint (A) for r_3; and $\pi_6 = \{r_1, r_3, r_5, r_7\}$ which violates one constraint (A) for r_1.

All of the preferred solutions are clearly more reasonable than the maximal solution because they do not allow five species $\{a, b, d, e, f\}$ to be created from no initial species at absolutely no cost!

Example 3. Consider the sporulation initiation network shown below, which is recreated from [31, p.165] with 11 reactions (in the rounded box) omitted for clarity. If I contains 9 species (in the dotted boxes) with no producers, and if T contains the single species $Spo0AP$, then there is just one maximal solution corresponding to the equilibrium state where the circled reactions are active (in addition to the dummy reaction r_0). But, this state has two unfounded loops. The first loop involves reactions r_{16} and r_{17} (shown below). It is unfounded because neither $KinA$ nor $KinAP$ can be present without the other already being present. The second loop involves reactions r_1 and r_3 (not shown). By contrast, there are two preferred solutions in which $\{r_0, r_4, r_{13}, r_{15}, r_{22}, r_{23}, r_{24}\}$ are active. In one case r_{20} is inactive, violating one constraint (C) for $Spo0AP$. In the other case r_{20} is also active, violating one constraint (A) for r_{20}.

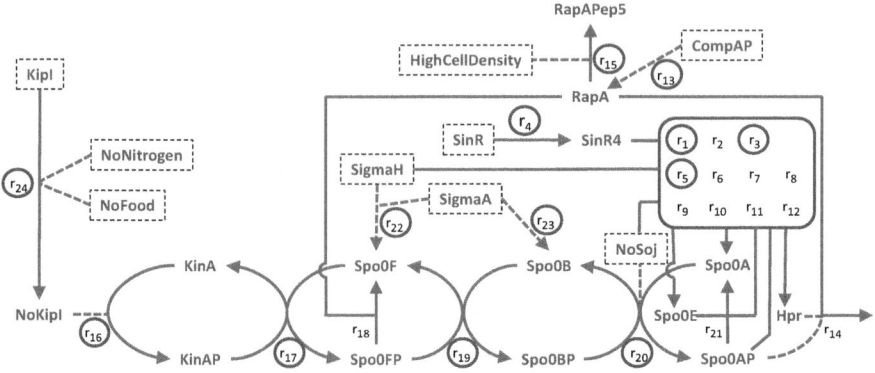

Example 4. Consider the MAPk network below which is recreated from [31, p.166]. If I contains 9 species (in the dotted boxes) with no producers, then there are two equilibrium states where $\{r_1\}$ or $\{r_1, r_3, r_4\}$ are active (along with the dummy reaction r_0). These are both maximal solutions. But, the latter has an unfounded loop involving two reactions r_3 and r_4 — since neither Raf nor Raf^* can be present without the other already being present. By contrast, there is just one preferred solution which corresponds to the first equilibrium state.

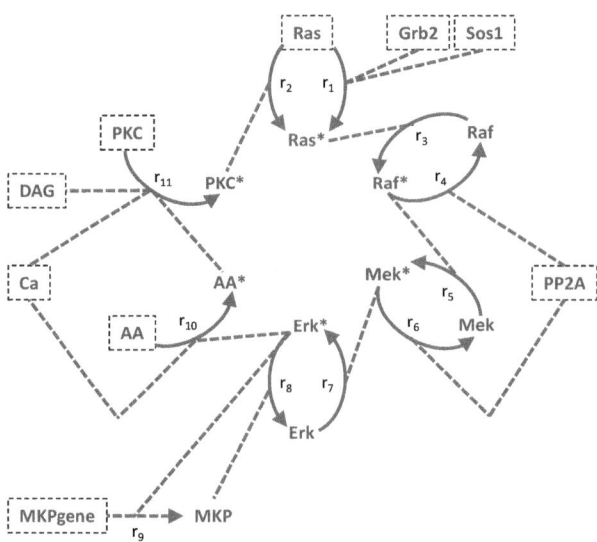

The existence of *multiple* cycles in both of these small case studies shows the practical importance of developing methods which handle them correctly.

8   ASP Computation of Maximal and Preferred Solutions

The difference between the maximal solutions of Tiwari et al. and the preferred solutions of the previous section is closely related to the distinction between the supported models and stable models of a logic program. For this reason, it is possible to translate the weighted Boolean constraints formalised above into a dual answer set program whose minimal-cost supported and stable models correspond, respectively, to the maximal and preferred network states.

Therefore modern ASP solvers which have the option of computing stable or supported models, can be used to find both maximal solutions under the original weights proposed by Tiwari et al. and preferred solutions under the corrected weights proposed in the previous section. It should be noted the ASP solver does not compute the true weight by naively applying the above correction to the original weight, but by ruling out from the very beginning any unfounded loops for which an appropriate penalty has not been paid.

Figure 1 shows how the network instance of Example 1 is easily encoded in the language of ASP. The first three lines simply list the available species along with any initial, target and/or forbidden species. The remaining lines each define one reaction in the network by explicitly listing the individual reactants, modifiers, and products of those reactions.

Figure 2 shows how the initialisation and weight constants of Example 1 are easily encoded in the language of ASP. The first two lines provide an automatic way of adding a dummy reaction r_0 that produces all initial species. The other lines specify the weights $x_A - x_F$ using the fact that $W = 9$, $k = 0$, and there is no preference for turning reactions on or off.

Figure 3 shows how the task of finding preferred or maximal states can be represented in the language of ASP. First, variables $R, R1, R2$ and S are defined over reactions and species. Then, three auxiliary predicates are defined. An *input* of a reaction is any species that is a reactant or modifier of that reaction. Two reactions *compete* if some species is an input of both and a reactant of one. A reaction is *viable* if all its inputs are present and all its competitors are inactive.

A reaction is active if it is viable and we choose not to pay a penalty that would allow us to render it inactive. Otherwise a reaction is active if it is not viable but we choose to pay a penalty that allows us to render it active anyway. These two penalties are both associated with weight x_A as each of them represents one way of violating constraint (A) in Definition 9.

A species is present if it is produced by an active reaction. Or a species is present if it is not produced by any reaction but we choose to pay a penalty that allows us to make it present anyway. This penalty is associated with weight x_B as it is equivalent to violating constraint (B) in Definition 9. Since we have explicit atoms denoting the presence of species, we need not add dummy completion reactions to the network.

In the cases discussed so far, the constraint literals indicate there is a choice between paying the penalty in order to violate a constraint or not violating the constraint in order not to pay the penalty.

```
species(a;b;c;d;e;f).
initial(a).
target(e).
reaction(r1). reactant(a,r1). product(b,r1).
reaction(r2). reactant(b,r2). product(a,r2). product(f,r2).
reaction(r3). reactant(b,r3). product(c,r3).
reaction(r4). reactant(c,r4). product(b,r4).
reaction(r5). reactant(c,r5). product(d,r5).
reaction(r6). reactant(d,r6). product(c,r6).
reaction(r7). reactant(d,r7). product(e,r7).
reaction(r8). reactant(e,r8). product(d,r8). modifier(f,r8).
```

Fig. 1. ASP encoding of network instance from Example 1

```
reaction(r0) :- initial(S).
product(S,r0) :- initial(S).
#const xA = 9.
#const xB = 9.
#const xC = 9.
#const xD = 999.
#const xE = 0.
#const xF = 0.
```

Fig. 2. ASP encoding of network initialisation and weights

The remaining penalties associated with not producing a target species x_C, and producing a forbidden species x_D, for turning on reactions x_E or turning off reactions x_F are deterministically applied.

As well as providing a semantics that correctly handles unfounded loops in reaction networks with cycles, another important advantage of the ASP framework proposed here is that it allows both numeric and logical constraints to be modularly added or removed. Such constraints can exploit the full expressive power of ASP in order to define new preferences in addition to or instead of the weight-based scheme proposed by Tiwari et al.

```
#domain species(S).
#domain reaction(R;R1;R2).

input(S,R) :- 1{reactant(S,R),modifier(S,R)}.
compete(R1,R2) :- 2{input(S,R1;R2)}, 1{reactant(S,R1;R2)}, R1≠R2.
viable(R) :- present(S0):input(S0,R), not active(R0):compete(R0,R).

active(R) :- viable(R), not pay("viable-reaction-inactive",R).
active(R) :- pay("unviable-reaction-active",R).

present(S) :- product(S,R), active(R).
present(S) :- pay("unsynthesisable-species-present",S).

{pay("unviable-reaction-active",R)} :-
                         not pay("viable-reaction-inactive",R).
{pay("viable-reaction-inactive",R)} :-
                         viable(R).
{pay("unsynthesisable-species-present",S)} :-
                         not product(S,R0):reaction(R0).

pay("target-species-absent",S) :- target(S), not present(S).
pay("forbidden-species-present",S) :- forbidden(S), present(S).
pay("arbitrary-reaction-inactive",R) :- not active(R)
pay("arbitrary-reaction-active",R) :- active(R).

#minimize
        [
        pay("unviable-reaction-active",R0):reaction(R0)=x_A ,
        pay("viable-reaction-inactive",R0):reaction(R0)=x_A ,
        pay("unsynthesisable-species-present",S0):species(S0)=x_B ,
        pay("target-species-absent",S0):species(S0)=x_C ,
        pay("forbidden-species-present",S0):species(S0)=x_D ,
        pay("arbitrary-reaction-inactive",R0):reaction(R0)=x_E ,
        pay("arbitrary-reaction-active",R0):reaction(R0)=x_F
        ].
```

Fig. 3. ASP encoding of maximal or preferred solutions

The set of all maximal or preferred solutions can be computed using the ASP system Gringo/Clasp [12,11]. All preferred solutions can be computed with the option `--opt-all` while maximal solutions can be computed with the options `--opt-all` and `--supp-models`. For all the examples in this paper, it took less than 200 ms to compute all minimal-cost solutions using Gringo 2.0.3 and Clasp 1.3.0 on a 2GHz Intel Pentium Duo laptop with 4Mb RAM.

9   Related Work

The key problem addressed in this paper is that of integrating numerical and logical information in order to better rank the solutions produced by biological network analysis. This appears to be an important problem because empirical evidence suggests that many solutions are usually produced in realistic problems [19,18].

Following Tiwari et al., we add numerical weights into a logical framework in order to define a preference on the solution space. The main difference is that our approach correctly handles networks with loops while theirs does not. Earlier work has approached the problem of cycles in other ways. For example, Tamura et al. [30] describe a pre-processing method to break cycles using feedback vertex sets; while Beasley and Planes [4] describe a post-processing method to check if solutions include cycles. By contrast, our method exploits the stability of ASP to appropriately penalise any unfounded loops.

Many authors have applied ASP to the analysis of biological networks. For example, Schaub and Thiele use ASP methods to address the task of computing metabolic scope and inverse scope [28]; while Dworschak et al. use a reduction of action languages to ASP to reason about biological networks [7]. Ray et al. have utilised an ASP system to propose revisions to metabolic networks in response to biological data generated by a Robot Scientist [26].

Much previous work in the analysis of metabolic networks employs heuristics aimed at maximising or minimising the number of reactions in a solution. For example, in metabolic pathway recovery, Beasley and Planes suggest minimising the number of reactions that must be added to a network [4]; while in the analysis of structural robustness, Tamura et al. suggest minimising the number of reactions that must be removed from a network [30]. Like Tiwari et al., our approach naturally includes both these possibilities as special cases by setting the preference parameter to 'on' or 'off'.

In principle, all constraint-based approaches with native numeric support, such as MaxSat, Integer Programming and ASP, allow the modular addition of constraints (although a potential advantage of ASP is that it more naturally allows those constraints to be expressed in terms of higher level logical concepts relevant to the domain of interest).

A complementary approach is proposed by Inoue et al. [14] who combine a first-order logical hypothesis-finding method with a probabilistic ranking based on Expectation Maximization (EM). Variants of this ranking technique could potentially be applied to other logical approaches, such as those which apply action languages to metabolic networks [7] or signalling pathways [2].

The approach presented in this paper focuses on the computation of stable states by abstracting away kinetic and stoichiometric parameters. There is a growing body of related work based on qualitative reasoning [16] which tries to approximate the solution of differential equations by reasoning over classes of such parameter values (so the parameters can still be used to some extent even if their precise values are unknown). Many other methods are explicitly concerned with the temporal evolution of biological systems, such as BIOCHAM, which answers queries expressed in a temporal logic [10].

10 Conclusion

This paper considered the integration of constraint-based and weight-based inference to find the most likely states of a reaction network with respect to given constraints. We showed how an existing state-of-the-art approach overestimates the weight of unreachable states in networks with cycles. We defined a corrected weight for penalising unfounded cycles and showed the impact of this on the sporulation and MAPk case studies. We showed how both methods can be implemented using an ASP solver. But it remains to assess the scalability of proposed approach and to evaluate its utility in real applications.

Notes

[1] Tiwari et al. show how constraints of the form (A) can be 'broken up' into finer constraints which allow the assignment of different weights which capture, for example, the fact that competition over a species which is consumed by two reactions is intuitively stronger than the competition over a species which is only consumed by one. Our approach can also be also refined in this way with no difficulties.

[2] Formally, an equilibrium state of an instance $\langle N, I, F, T \rangle$ over a set of species S is equivalent to any state of the extended network $N \cup \{\langle \emptyset, \emptyset, I \rangle\} \cup \{\langle \emptyset, \emptyset, \{s\}\rangle \mid s \in S/I\}$ which satisfies constraints (A-D). Note the definitions imply that the network is first initialised and then completed.

[3] Tiwari et al. define a semantics for their reaction networks in terms of nondeterministic state transition systems [31, p.158] and associate equilibrium states with fixpoints of such transition systems. But a key drawback of their semantics is that it does not distinguish between fixpoints that are reachable from the initial state and fixpoints that are not. Fundamentally, this is the root cause underlying the counterintuitive properties of their approach that we address in this paper.

Acknowledgements. Thanks to the anonymous reviewers for their useful feedback. Thanks to Asish Tiwari for his helpful comments. Thanks to Research Councils UK for funding the first author through a research fellowship in Exabyte Informatics.

References

1. Alsinet, T., Manyà, F., Planes, J.: An efficient solver for weighted max-sat. Journal of Global Optimization 41, 61–73 (2008)
2. Baral, C., Chancellor, K., Tran, N., Tran, N.L., Joy, A., Berens, M.: A knowledge based approach for representing and reasoning about signaling networks. In: Proc. 12th Int. Conf. on Intelligent Systems for Molecular Biology, pp. 15–22 (2004)
3. Bay, S., Shrager, J., Pohorille, A., Langley, P.: Revising regulatory networks: From expression data to linear causal models. Journal of Biomedical Informatics 35, 289–297 (2003)
4. Beasley, J., Planes, F.: Recovering metabolic pathways via optimization. Bioinformatics 23(1), 92–98 (2007)
5. Bodei, C., Bracciali, A., Chiarugi, D.: On deducing causality in metabolic networks. BMC Bioinformatics 9(4) (2008)
6. Clark, K.: Negation as failure rule. In: Gallaire, H., Minker, J. (eds.) Logic and Data Bases, pp. 293–322. Plenum Press (1978)
7. Dworschak, S., Grell, S., Nikiforova, V., Schaub, T., Selbig, J.: Modeling Biological Networks by Action Languages via Answer Set Programming. Constraints 13(1-2), 21–65 (2008)
8. Eker, S., Knapp, M., Laderoute, K., Lincoln, P., Talcott, C.: Pathway Logic: Executable models of biological networks. In: Proc. 4th Int. Workshop on Rewriting Logic and Its Applications (2002)
9. Fages, F.: A new fixpoint semantics for general logic programs compared with the well-supported and stable model semantics. New Generation Computing 9, 425–443 (1991)
10. Fages, F., Soliman, S., Chabrier-Rivier, N.: Modelling and querying interaction networks in the biochemical abstract machine BIOCHAM. Journal of Biological Physics and Chemistry 4, 64–73 (2004)
11. Gebser, M., Kaufmann, B., Neumann, A., Schaub, T.: *clasp*: A Conflict-Driven Answer Set Solver. In: Baral, C., Brewka, G., Schlipf, J. (eds.) LPNMR 2007. LNCS (LNAI), vol. 4483, pp. 260–265. Springer, Heidelberg (2007)
12. Gebser, M., Schaub, T., Thiele, S.: GrinGo: A New Grounder for Answer Set Programming. In: Baral, C., Brewka, G., Schlipf, J. (eds.) LPNMR 2007. LNCS (LNAI), vol. 4483, pp. 266–271. Springer, Heidelberg (2007)
13. Gelfond, M.: Answer sets. In: Handbook of Knowledge Representation, pp. 285–316. Elsevier (2007)
14. Inoue, K., Sato, T., Ishihata, M., Kameya, Y., Nabeshima, H.: Evaluating Abductive Hypotheses using an EM Algorithm on BDDs. In: Proceedings of the 21st International Joint Conference on Artificial Intelligence, pp. 810–815 (2009)
15. Juvan, P., Demsar, J., Shaulsky, G., Zupan, B.: GenePath: from mutations to genetic networks and back. Nucleic Acids Research 33 (2005)
16. King, R., Garrett, S., Coghill, G.: On the use of qualitative reasoning to simulate and identify metabolic pathways. Bioinformatics 21(9), 2017–2026 (2005)
17. King, R., Whelan, K., Jones, F., Reiser, P., Bryant, C., Muggleton, S., Kell, D., Oliver, S.: Functional Genomic Hypothesis Generation and Experimentation by a Robot Scientist. Nature 427, 247–252 (2004)
18. Klamt, S., Stelling, J.: Combinatorial complexity of pathway analysis in metabolic networks. Molecular Biology Reports 29(1-2), 233–236 (2002)
19. Küffner, R., Zimmer, R., Lengauer, T.: Pathway analysis in metabolic databases via differetial metabolic display (DMD). In: German Conference on Bioinformatics, pp. 141–147 (1999)

20. Kügel, A.: Improved exact solver for the weighted max-sat problem. In: Proc. of the 2010 Pragmatics of SAT Workshop (2010)
21. Lifschitz, V.: What is answer set programming? In: Proc. 23rd AAAI National Conf. on Artificial Intelligence, pp. 1594–1597. AAAI Press (2008)
22. Lloyd, J.: Foundations of Logic Programming. Springer, Heidelberg (1987)
23. Marek, W., Subrahmanian, V.S.: The relationship between stable, supported, default and autoepistemic semantics for general logic programs. Theoretical Computer Science 103, 365–386 (1992)
24. Muggleton, S., King, R., Sternberg, M.: Protein secondary structure prediction using logic-based machine learning. Protein Engineering 5(7), 647–657 (1992)
25. Papatheodorou, I., Kakas, A., Sergot, M.: Inference of Gene Relations from Microarray Data by Abduction. In: Baral, C., Greco, G., Leone, N., Terracina, G. (eds.) LPNMR 2005. LNCS (LNAI), vol. 3662, pp. 389–393. Springer, Heidelberg (2005)
26. Ray, O., Whelan, K., King, R.: Automatic Revision of Metabolic Networks through Logical Analysis of Experimental Data. In: De Raedt, L. (ed.) ILP 2009. LNCS (LNAI), vol. 5989, pp. 194–201. Springer, Heidelberg (2010)
27. Srinivasan, A., Muggleton, S., Sternberg, M., King, R.: Theories for Mutagenicity: A Study in First-Order and Feature-Based Induction. Journal of Artificial Intelligence 85(1-2), 277–299 (1996)
28. Schaub, T., Thiele, S.: Metabolic Network Expansion with Answer Set Programming. In: Hill, P.M., Warren, D.S. (eds.) ICLP 2009. LNCS, vol. 5649, pp. 312–326. Springer, Heidelberg (2009)
29. Tamaddoni-Nezhad, A., Chaleil, R., Kakas, A., Sternberg, M., Nicholson, J., Muggleton, S.: Modeling the effects of toxins in metabolic networks. IEEE Engineering in Medicine and Biology 26, 37–46 (2007)
30. Tamura, T., Takemoto, K., Akutsu, T.: Measuring Structural Robustness of Metabolic Networks under a Boolean Model Using Integer Programming and Feedback Vertex Sets. In: Proc. 3rd Int. Conf. on Complex, Intelligent and Software Intensive Systems, pp. 819–824. IEEE (2009)
31. Tiwari, A., Talcott, C., Knapp, M., Lincoln, P., Laderoute, K.: Analyzing Pathways Using SAT-Based Approaches. In: Anai, H., Horimoto, K., Kutsia, T. (eds.) AB 2007. LNCS, vol. 4545, pp. 155–169. Springer, Heidelberg (2007)

Author Index